Detecting and Investigating Environmental Crime

Petter Gottschalk

Detecting and Investigating Environmental Crime

The Case of Tjøme Island

palgrave
macmillan

Petter Gottschalk
Department of Leadership
and Organizational Behavior
BI Norwegian Business School
Oslo, Norway

ISBN 978-3-030-74183-9 ISBN 978-3-030-74184-6 (eBook)
https://doi.org/10.1007/978-3-030-74184-6

Cover credit: guvendemir

This Palgrave Macmillan imprint is published by the registered company Springer Nature Switzerland AG
The registered company address is: Gewerbestrasse 11, 6330 Cham, Switzerland

Contents

List of Figures

List of Tables

1

Introduction

Environmental harm and crime have received increased attention in recent years (Böhm, 2020; Gibbs & Boratto, 2017; Huisman & Erp, 2013; Lynch, 2020). Environmental crime is examined in this book in the perspective of detection of offenders and investigation of offenses. The main environmental crime study presented in this book is concerned with the case of the Tjøme Island in Norway. The rich and mighty have built summerhouses and vacation resorts for themselves on the shoreline that has harmed wildlife and public access. The unique nature and climate along the seaside on the island and around the island is for residents, holidaymakers, visitors, and wildlife. An enthusiastic whistleblower started to review municipal documents and found strange pieces of information from an architect, an internal manager, external attorneys, and summerhouse owners. Local police was reluctant to investigate the matter, while national police with expertise on environmental crime initiated investigations that lead to criminal prosecution (Aunemo, 2020; Berglihn, 2020).

Before the main case of Tjøme Island is introduced in this book, the case study of Trafigura at Vest Tank in Chapter 3 applies the structural model of convenience theory in Fig. 2.1 to examine motive, opportunity,

P. Gottschalk, *Detecting and Investigating Environmental Crime*,
https://doi.org/10.1007/978-3-030-74184-6_1

and willingness. There was a threat to the corporation of not getting rid of the dangerous waste that caused strain, pain, and uncertainty (Knudssøn & Bakke, 2009; Langton & Piquero, 2007). Surprisingly often, environmental crime and other forms of corporate crime are not the result of the well-known phenomenon of greed (Goldstraw-White, 2012). Rather, avoidance of corporate problems is more common (Alalehto, 2020; Blickle et al., 2006), as well adaption to criminal market forces (Leonard & Weber, 1970; Schoultz & Flyghed, 2020a, 2020b), and required cooperation with criminal networks and cartels (Goncharov & Peter, 2019).

Increased profit for business is sometimes at the expense of environmental degradation. The cost of this degradation is borne by society as a whole and not the business per se.

White-collar offenses, such as environmental crime, have typically three main properties—(1) special access, (2) spatial separation, and (3) superficial appearance of legitimacy—that Benson and Simpson (2018: 132) found evidence in Woburn in the United States:

> All these characteristics are illustrated in the tragic 1980s case of the W.R. Grace Corporation and a small working-class town in Massachusetts. The case against the W.R. Grace Corporation began in the mid-1970s when the residents of Woburn, Massachusetts, started to realize that an abnormally large number of local children had contracted leukemia.

Toxic waste from the company had filtered into the groundwater and from there into the homes and bodies of the people of Woburn. After a few years of legal maneuvering, the company settled with the families. The company also pled guilty of the crime of providing false statements to the Environmental Protection Agency (EPA) about its toxic waste disposal practices.

In the United States, the EPA has criminal investigative and civil enforcement powers. Environmental crime is nevertheless convenient, since very few offenders ever become prosecuted (Benson & Simpson, 2018). The Norwegian agency 'Miljødirektoratet' for environmental protection has no such EPA powers. Therefore, the Norwegian authority for investigation and prosecution of economic and environmental crime

('Økokrim')—which is similar to the serious fraud office (SFO) in the United Kingdom—handles a few cases every year. In the United States, Norway and elsewhere, criminal opportunities tie themselves to enforcement practices, which are very limited.

After the Deepwater Horizon oil spill in the Gulf of Mexico in April 2010, prosecutors brought criminal charges against four British Petroleum (BP) executives, but no one ended up in prison (Freeh, 2013; Thompson, 2017). The oil spill that occurred when the oil rig Deepwater Horizon exploded and sank into the Gulf of Mexico, EPA officials considered the largest spill in U.S. history. The explosion killed 11 rig workers and injured many others, and the pipe leaked oil and gas on the ocean floor for 87 days until the Macondo well was capped (Fowler, 2014; Kang & Thosuwanchot, 2017; Thompson, 2017). In the interim, an estimated 4.9 million barrels of oil leaked into the Gulf, resulting in substantial deterioration of the ecosystem (Benson & Simpson, 2018).

The primary corporate actors in this disaster were oil company BP, rig company Transocean, and service company Halliburton. BP pleaded guilty to criminal charges, admitting 11 counts of felony manslaughter, obstruction of Congress, and a series of environmental crime, and agreed to a $4 billion fine. BP settled class action claims by individuals and businesses that experienced serious harm economically by the spill with $11 billion. The oil company later challenged the settlement with the help of fraud examiner Freeh (2013) but lost. In 2015, BP agreed to pay $18.7 billion to settle all remaining federal and state claims (Benson & Simpson, 2018).

Four BP individuals were charged in the Deepwater Horizon oil spill case: Kurt Mix (chief engineer), Robert Kaluza (rig supervisor), Donald Vidrine (rig supervisor), and David Rainey (vice president of exploration). Prosecutors brought criminal charges ranging from obstruction of justice (deleting e-mails and voice mail communications) and lying to Congress, to seaman's manslaughter and involuntary manslaughter. Seaman's manslaughter is a special type of manslaughter charge in which criminal penalties are relevant when loss of life results from a wrongful act related to misconduct, negligence, or inattention to duties. The consequence of all these court proceedings is that no one would go to prison for the BP disaster (Benson & Simpson, 2018).

Prosecution of environmental crime typically ends up like the BP pollution scandal, where the company suffers heavy financial losses, but no responsible managers have to go to prison. Moreover, as suggested by Benson and Simpson (2018), businesses have a great deal of influence over the nature of the oversight and governance that they have to adhere to, thereby mostly avoiding detection and prosecution.

In the perspective of convenience theory, environmental crime has financial rewards that white-collar offenders can gain from the offense. There is rationality in saving money that the company would normally spend on environmentally sound disposal of toxic waste, and that outweigh its costs of low risk of detection. The benefit of crime outweighs its costs.

Environmental crime by white-collar offenders displays all of the characteristic features of white-collar crime suggested by Benson and Simpson (2018). The company has a legitimate right to do what the organization does. Thus, it has specialized and legitimate access to crime resources. Those who will eventually suffer harm may be at a distance from where the offense occurs. Finally, because there often are no obvious signs that wrongdoing has happened, the company appears to be doing nothing wrong, at least as observable from the outside. The reason for detection of wrongdoing at Vest Tank in Norway was an explosion that occurred, and people living nearby suffered from the air pollution.

Chapter 2 introduces the theoretical framework applied in this book. The theory of convenience suggests that the tendency to commit white-collar crime in terms of environmental crime is dependent on economic possibilities and threats, professional opportunities to commit and conceal wrongdoing, as well as the extent of personal willingness for deviant behavior.

Corporate environmental crime cases such Grace Corporation and British Petroleum are interesting. New corporate crime cases emerge all the time. The case of Vest Tank in Norway in Chapter 3 is only one of them. The research literature has recently focused on such cases in criminology, management, and law.

Therefore, this book focuses on a different kind of environmental crime where individuals, rather than corporations, are the offenders. Specifically, the book discusses various aspects of the Tjøme case in

Norway. The purpose of the case study is to illustrate the many forces and aspects that cause environmental crime. The purpose is also to describe the process of police investigation from detection to prosecution in court, where individuals change their position from suspected to charged and indicted, and eventually convicted persons.

Chapter 4 presents the Tjøme Island case study. The island deserves comparison to Cape Cod in Massachusetts in the United States, which is an attractive resort area for summer vacation. The island of Tjøme is an extremely attractive resort where wealthy individuals started to build small cabins, while they now build mansions and palaces as summer-houses and holiday homes. There is a strict regulation that nobody can build anything in the hundred-meter belt along the shoreline on the island. The beach zone belt is for public access and wildlife. The unique nature and climate along the seaside on the island around the island is for residents, holidaymakers, and visitors. The island is located south of Oslo, the capital of Norway. Most of the holidaymakers and visitors on the island are from Oslo. It takes them two hours to drive their cars from the city of Oslo via the city of Tønsberg to the island.

While the ban on construction work along the shoreline is absolute, the local municipality has issued surprisingly many dispensations in recent years. In addition, surprisingly many rich and mighty individuals have ignored the construction ban. Summerhouse owners have expanded their buildings, acquired swimming pools and tennis courts, built docs for their boats, and they turned areas between the buildings and the water into private parks. Public access has thus disappeared most places, and the landscape has changed.

Two cabin owners who did not change their summerhouse noticed the fundamental changes on the island in recent years. They asked themselves how this could happen. They started to review municipal files on building permits. After a short while, they found that low-level officials had granted building permits in the municipality and never involved political bodies. Furthermore, they noticed that applicants typically first applied for approval of marginal changes that low-level municipal officials approved. Then after a short while, applicants returned with further marginal changes that the officials also approved. In some cases, the two cabin owners detected systematic series of change applications for summerhouse expansions to mansions and palaces with tennis courts and swimming pools, in addition to large terraces, parks, and docs.

The two cabin owners found that the incremental applications did not arrive directly from the rich and mighty people from Oslo. A local architectural firm issued all the applications, and all applications from that specific architect people always approved in the municipality. Furthermore, some of the rich and mighty people had set up the local architectural firm. The chief architect in the firm that carried his name had strong links to municipal employees handling building permits. One of them left the municipal position temporarily to work for the firm, where he designed applications for rich people. When he returned to his municipal position, he approved applications from the firm that he had designed. There are many more elements to this story, which we return to in this book.

One group of professionals that we cover in this book are attorneys who work for the rich and powerful people from Oslo. If the rich and powerful do not get what they want, they hire the best attorneys to get what they want. If the rich and powerful experience detection and face threats of sanctions, then they hire the best attorneys to avoid the sanctions. The attorneys apply substance defense, as well as information control and symbol defense on behalf of their clients.

Chapter 5 introduced white-collar attorneys who help clients avoid the criminal justice system. Attorneys help rich and mighty people to get what they want on the shoreline. Attorneys find convenient loopholes for mansions, and they find various ways of making it problematic for municipal officials to implement rules and regulations.

Chapter 5 describes the three strategies of prominent attorneys that the Yale studies introduced a few decades ago. First, attorneys apply substance defense where they have expert knowledge of laws, regulations, and court rulings. Next, attorneys apply information control where they protect information that is to the disadvantage for their clients, while they willingly supply information that is advantageous for their clients. Finally, attorneys apply symbol defense where they attempt to portray their clients as sympathetic individuals who have become victims of municipal interference.

Policing environmental crime is concerned with collection of information that can enable decisions whether or not someone has violated the law. Police investigation is a goal-driven and rule-restricted activity that the prosecuting authority in Norwegian police is formally heading. Police officers carry out the practical investigation methods, while the prosecuting branch in the police is in charge of everything involved with decisions about coercive measures and legal proceedings, such as the taking of evidence and evidence statements and legal claims abroad. The aim of the investigation is to elucidate whether potential offenders have transgressed various penal provisions. A professional investigation by the police should be objective. A professional investigation against a suspect will cover both what speaks in favor of the person and what speaks in disfavor of the person in question. Investigators should act as impartial as humanly possible. Knowledge and awareness of typical human error sources such as confirmation traps, tunnel vision and social stereotypes are important. The same is the need for a knowledge-based and transparent examination methodology. An investigation shall examine whether something has happened or will happen, and in that case what, when, where, how, why, who has participated, and what role they may have played.

The two cabin owners turned into whistleblowers and became important sources of information for the police. Chapter 7 tells what information the whistleblowers as informants initially provided police investigators to help them get started in the first year of investigation. Chapters 8 and 9 tell how the police investigation evolved in the two following years. The three Chapters 7–9 on the police investigation steps in 2018, 2019, and 2020 offer valuable insights into the information collection and the importance of the whistleblowers as information providers in the field. We close the case description before court hearings that the district court had scheduled for January 2021.

The purpose of the thorough review of the Tjøme case generally and the police investigation specifically is to illustrate the various aspects of environmental crime by rich and mighty individuals. This book thus opens up a new avenue for researchers who want to study individual wrongdoing rather than corporate wrongdoing when it comes to environmental crime. The interplay between potentially corrupt public officials, professionals like architects and attorneys, and rich and mighty individuals is an interesting and challenging arena for law enforcement.

Chapter 10 introduces some general perspectives on environmental crime. Punishing environmental crime is dependent on public perceptions of crime seriousness. While sanctioning by fines was common for both individual and corporate wrongdoing, prison sentences have become more relevant in recent years. The perspective of green criminology has contributed to the perceived seriousness of environmental crime where class and inequality are relevant concerns. The perspective of green criminology politics emphasizes that in the long run, law enforcement has to stay consistent with the priorities of the citizens in general and the ruling class in particular. The perspective of green criminology opens up several avenues for research.

Chapter 11 discusses private policing of white-collar crime. Private policing has been a controversial issue for quite some time. Fraud examiners from global auditing firms and local law firms conduct internal investigations, resulting in reports of investigations that are the clients' confidential property. The chapter presents four disclosed reports of investigations to illustrate the theory of convenience for white-collar crime and to determine the examination maturity in private policing. The suggested maturity model with five levels determines whether the investigation as chaotic or messy, or whether the investigation contributed to disclosure or clarification. The most mature level is investment, where the benefits of the examination exceed the costs of the examination. Many fraud examiners have a long way to go before they deserve the label of professional investigators or examiners. Future research may help professionalize the business of private policing of economic crime. The four reports of investigations are from Moldova, the Netherlands, Sweden, and Switzerland. None of the reports are concerned with environmental crime, as no such reports have been possible to identify and retrieve.

Chapter 12 presents a case study of traditional white-collar crime by a self-proclaimed environmentalist in Sweden. The case study illustrates how the structural model of convenience theory can be applied to a specific white-collar offender to identify relevant convenience themes that can explain the occurrence of crime.

This book is important, as "it is still necessary to legitimize the study of green crimes, given criminology's obsession with street crimes" (Lynch, 2020).

This book uses several crime terms that are interconnected: environmental crime, white-collar crime, and green crime. Both environmental crime and green crime are committed by white-collar offenders. Furthermore, green crime refers to crime committed against the environment. However, the terms are not interchangeable as they have different emphasis. White-collar crime emphasizes the privileged role of the offender who has access to resources to commit and conceal crime, and to resources to defend wrongdoing by means of specialized and well-paid attorneys. Environmental crime emphasizes the harm suffered by wildlife, biodiversity, and natural resources. Green crime emphasizes laws and regulations to protect the environment and to prevent damage and harm to health and other problems when the land, water, or air that living organisms interact with is polluted, damaged, or destroyed. The different terms are applied in this book according to the reviewed literature on the various topics.

However, corporate crime is different from occupational crime. Corporate crime sometimes labeled organizational offending is resulting from offenses benefitting the organization. Occupational crime benefits individuals. For both types, this book takes the approach that people must be held personally responsible to make any significant enforcement impact.

Bibliography

Alalehto, T. (2020). Corporate crime: A logical misconception, but with one analytical point, *Journal of Financial Crime*, published online https://doi.org/10.1108/jfc-04-2020-0063.

Aunemo, E. W. (2020). Advokaten til millionærarving Celestine (22) krever at skattebetalerne skal betale honoraret hennes (The millionaire heiress attorney demands that the taxpayers pay her fees), daily local Norwegian newspaper *Tønsbergs Blad*. www.tb.no, published January 19.

Benson, M. L., & Simpson, S. S. (2018). *White-collar crime: An opportunity perspective* (3rd ed.). Routledge.

Berglihn, H. (2020). Hanne Madsen må i fengsel: -Viktig at det reageres strengt (Hanne Madsen must go to jail: -It is important that the reaction is strict), daily Norwegian business newspaper *Dagens Næringsliv*. www.dn.no, published June 29.

Blickle, G., Schlegel, A., Fassbender, P., & Klein, U. (2006). Some personality correlates of business white-collar crime. *Applied Psychology: an International Review, 55*(2), 220–233.

Böhm, M. L. (2020). Criminal business relationships between commodity regions and industrialized countries: The hard road from raw material to new technology. *Journal of White Collar and Corporate Crime, 1*(1), 34–49.

Fowler, T. (2014, January 27). BP's new tactic in oil spill claims: Go after the 'special master'. *The Wall Street Journal*. www.wsj.com.

Freeh, L. J. (2013, September 6). *Independent external investigation of the Deepwater Horizon court supervised settlement program* (Report of Special Master Louis J. Freeh). www.laed.uscourts.gov.

Gibbs, C., & Boratto, R. (2017). Environmental crime. In H. N. Pontell (Ed.), *Oxford encyclopedia of criminology and criminal justice*, published online https://oxfordre.com/criminology.

Goldstraw-White, J. (2012). *White-collar crime: Accounts of offending behavior*. Palgrave Macmillan.

Goncharov, I., & Peter, C. D. (2019). Does reporting transparency affect industry coordination? Evidence from the duration of international cartels. *The Accounting Review, 94*(3), 149–175.

Huisman, W., & Erp, J. (2013). Opportunities for environmental crime. *British Journal of Criminology, 53*, 1178–1200.

Kang, E., & Thosuwanchot, N. (2017). An application of Durkheim's four categories of suicide to organizational crimes. *Deviant Behavior, 38*(5), 493–513.

Knudssøn, K., & Bakke, S. (2009). A spreading of toxic gags. *The Guardian*. www.theguardian.com, published October 26.

Langton, L., & Piquero, N. L. (2007). Can general strain theory explain white-collar crime? A preliminary investigation of the relationship between strain and select white-collar offenses. *Journal of Criminal Justice, 35*(1), 1–15.

Leonard, W. N., & Weber, M. G. (1970). Automakers and dealers: A study of criminogenic market forces. *Law & Society Review, 4*(3), 407–424.

Lynch, M. J. (2020). Green criminology and environmental crime: Criminology that matters in the age of global ecological collapse. *Journal of White Collar and Corporate Crime, 1*(1), 50–61.
Schoultz, I., & Flyghed, J. (2020a). Denials and confessions: An analysis of the temporalization of neutralizations of corporate crime. *International Journal of Law, Crime and Justice*, published online https://doi.org/10.1016/j.ijlcj.2020.100389.
Schoultz, I., & Flyghed, J. (2020b). "We have been thrown under the bus": Corporate versus individual defense mechanisms against transnational corporate bribery charges. *Journal of White Collar and Corporate Crime*, published online https://doi.org/10.1177/2631309x20911883.
Thompson, R. (2017, October 29). With BP settlement claims winding down, Lafayette lawyer Parick Juneau turns attention to Takata air bag recall. *The Advocate*. www.theadvocate.com.

2

Theory of Crime Convenience

The emerging theory of convenience for white-collar offenses, such as environmental crime, is concerned with economic possibilities and threats, professional opportunity to commit and conceal crime, as well as personal willingness for deviant behavior. The research literature on convenience theory is growing (e.g., Braaten & Vaughn, 2019; Chan & Gibbs, 2020; Dearden & Gottschalk, 2020; Gottschalk, 2017, 2019).

White-collar offenders commit financial crime in the course of their occupations or associated with their occupations as powerful elite members in society (Craig, 2019; Craig & Piquero, 2016; Dearden, 2016, 2017, 2019; Jordanoska, 2018; Klenowski et al., 2011; Onna, 2020; Onna & Denkers, 2019). Convenience theory suggests that the financial motive in white-collar crime is to explore possibilities and avoid threats; the organizational opportunity is to commit as well as conceal crime, while the willingness for deviant behavior derives from personal ability to justify actions and neutralize guilt feelings (Gottschalk, 2017, 2019). The theory of convenience is an integrated and deductive perspective based on the synthesis of individual-level, group-level, and nation-level themes (Chan & Gibbs, 2020). Convenience theory builds on previous theoretical perspectives on fraud, such as the fraud triangle,

P. Gottschalk, *Detecting and Investigating Environmental Crime*,
https://doi.org/10.1007/978-3-030-74184-6_2

the fraud scale, the fraud diamond, the MICE model, and the ABC model (Desai, 2020), as well as general theoretical perspectives on economic crime from criminology, sociology, psychology, management, and organizational behavior (Gottschalk, 2017, 2019).

The Concept of Convenience

Convenience is the state of being able to proceed with something with little effort or difficulty, avoiding pain and strain (Mai & Olsen, 2016). Convenience is savings in time and effort (Farquhar & Rowley, 2009), as well as avoidance of pain and obstacles (Higgins, 1997). Convenience is a relative concept concerned with efficiency in time and effort as well as reduction in pain and solution to problems (Engdahl, 2015). Convenience is an advantage in favor of a specific action to the detriment of alternative actions. White-collar offenders choose the most convenient path to reach their goals (Wikstrom et al., 2018).

White-collar crime results from delinquent behavior by individuals in competent positions (Piquero, 2018). White-collar offenders commit and conceal their crime in a professional setting (Yeok et al., 2020) where they have legitimate access to premises, resources, and systems (Logan et al., 2019). The benefit from white-collar crime might be financial gain, personal adventure or some other desired outcome (Craig & Piquero, 2017; Jordanoska, 2018; Sutherland, 1939, 1983; Williams et al., 2019).

Figure 2.1 illustrates the structure of white-collar convenience as derived from the research literature. The extent of white-collar crime convenience manifests itself by motive, opportunity, and willingness. The motive is either personal crime to benefit the individual or corporate crime to benefit the organization because of possibilities or threats. The ability of white-collar offenders to commit and conceal crime links to their privileged position, the social structure, and their orientation to legitimate and respectable careers (Friedrichs et al., 2018). The personal willingness for deviant behavior manifests itself by offender choice and perceived innocence. The choice of crime can be caused by deviant identity, rational consideration, or learning from others. Social identity is an individual's self-concept as an organizational and societal member

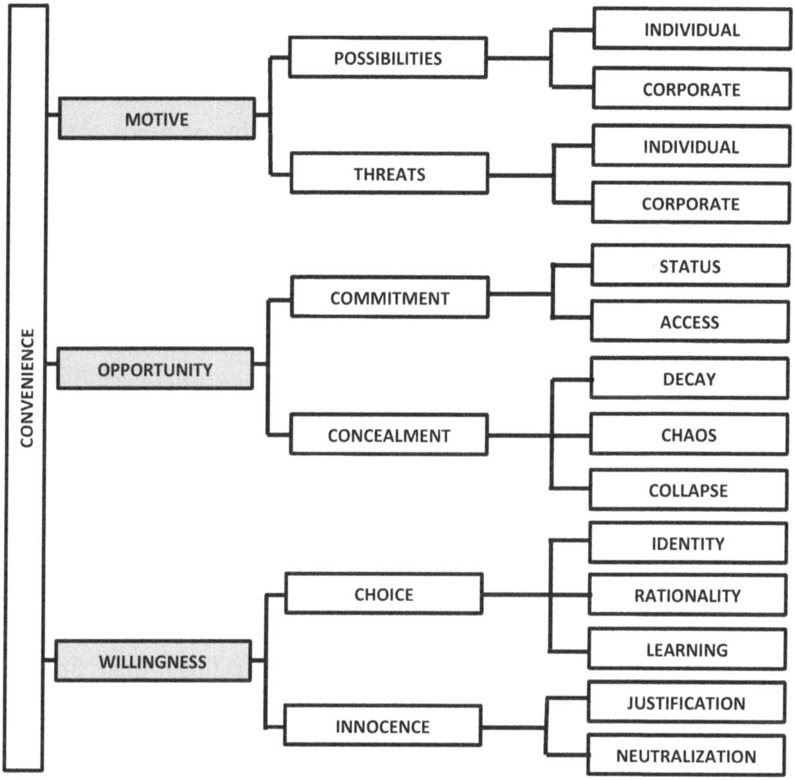

Fig. 2.1 Structural model of convenience theory

(Piening et al., 2020). The perceived innocence at crime manifests itself by justification and neutralization (Schoultz & Flyghed, 2020a, 2020b). Identity, rationality, learning, justification, neutralization, and lack of self-control all contribute to making white-collar crime action a convenient behavior for offenders (Craig & Piquero, 2017; Engdahl, 2015; Holtfreter et al., 2010; Sutherland, 1983; Sykes & Matza, 1957).

In criminology, the conceptualization of deviant behavior is implicitly negative actions that lack conformity to norms (Sykes & Matza, 1957). Deviant behavior typically causes harm to victims. Offenders depart from norms in destructive ways. In our context, deviance is negative departure from norms by white-collar offenders (Gottschalk, 2019).

Convenient Economic Motive

It is convenient to use financial resources to explore possibilities and avoid threats (Naylor, 2003). Climb the hierarchy of needs for status and success (Maslow, 1943), realize the American dream of prosperity (Schoepfer & Piquero, 2006), satisfy the need for acclaim as a narcissist (Chatterjee & Pollock, 2017), and restore the perception of equity and equality (Clark et al., 2010) are some of the perspectives integrated in the motive dimension of convenience theory. In addition, goal setting is a common practice in the field of organizational behavior, where high performance goals tend to encourage unethical behavior (Welsh et al., 2020). The extra profit from financial crime enables the offender to handle desired possibilities and potential threats. It is mainly the convenience of extra profit, rather than the convenience of illegal profit, that is important in the motive dimension of convenience theory. However, under certain circumstances, there might be some extra benefits from illegal extra profit rather than extra profit in general, since illegal funds avoid the attention of external and internal control mechanisms, including compliance functions (Kawasaki, 2020). Illegitimate financial gain can thus find its ways into exploring possibilities and avoiding threats that recorded funds cannot.

In the case of illegal construction and expansion of summerhouses on the shoreline on Tjøme Island, including swimming pools, tennis courts, and parks, the motive for environmental crime is the illegitimate gain in offenders' standard of living (Maslow, 1943). The rise in standard of living follows an accumulation of wealth (Schoepfer & Piquero, 2006), as summerhouses on the beach are a scarce resource of increased value. Environmental crime is profit-oriented avoidance of costs and access to possibilities that most people do not have. Some mansion owners have a motive of satisfying their need for acclaim as narcissists (Chatterjee & Pollock, 2017), where the mansion combined with luxury boats, tennis court, and swimming pool can mirror success and prestige. As members of the elite, some offenders have a motive of restoring the perception of equity and equality in the elite (Leigh et al., 2010). As members of a dynasty, some offenders have a motive of satisfying the desire to help others as social concern (Agnew, 2014).

Environmental criminals want to reach personal objectives that they believe justify means (Jonnergård et al., 2010). They may suffer from greed (Goldstraw-White, 2012) where nothing is ever enough in terms of comfort, exclusiveness, and grandiosity on the shoreline. Municipal officials, architects, attorneys, and rich heirs can enjoy mutual benefits in exchange relationships (Huang & Knight, 2017). The goal is often to maximize profits (Naylor, 2003) in terms of an exclusive vacation resort in the beach zone.

Environmental criminals want to avoid loss of self-esteem (Crosina & Pratt, 2019) after expansion failure, and they want to avoid pain, strain, and uncertainty (Langton & Piquero, 2007) related to personal goal achievement for vacation resort. They do not want to fall from position in the privileged elite (Blickle et al., 2006; Piquero, 2012) on the island. They join networks (Goncharov & Peter, 2019; Leonard & Weber, 1970; Schoultz & Flyghed, 2020a, 2020b) to get what they want.

Convenient Opportunity Structure

There is convenient access to resources to commit and conceal elite crime. Legitimate access to premises and systems (Benson & Simpson, 2018), specialized access in routine activity (Cohen & Felson, 1979), blame game by misleading attribution to others (Eberly et al. 2011), and institutional deterioration (Rodriguez et al., 2005) are some of the perspectives integrated in the opportunity dimension of convenience theory. A typical white-collar offender does not go into hiding as many street criminals do. Rather, the offender conceals financial crime among legal transactions to make illegal transactions seem legitimate, or the offender conceals financial crime by removing certain activities from the books. A typical white-collar offender who has convenient legitimate access to commit crime might spend most of the energy on concealing crime in the professional context (Huisman & Erp, 2013; McClanahan & South, 2020).

In the case of illegal construction and expansion of summerhouses on the shoreline on Tjøme Island, including swimming pools, tennis courts, and parks, the opportunity for environmental crime is potentially rooted in all convenience themes in Fig. 2.1. The status can derive from executive language that people do not understand (Ferraro et al., 2005), elite members too big to fail and too powerful to jail (Pontell et al., 2014), and blame game by misleading attribution to others (Eberly et al., 2011). Status can also derive from offender humor distraction from deviant behavior (Yam et al., 2018) and power inequality between the elite and others (Patel & Cooper, 2014). Power is relative control over outcomes through the capacity to withdraw rewards or introduce punishments (Berdahl, 2007).

Status is an individual's social rank within a formal or informal hierarchy, or the person's relative standing along a valued social dimension. Status is the extent to which an individual is respected and admired by others, and status is the outcome of a subjective assessment process (McClean et al., 2018). High-status individuals enjoy greater respect and deference from, as well as power and influence over, those who are positioned lower in the social hierarchy (Kakkar et al., 2020: 532):

> Status is a property that rests in the eyes of others and is conferred to individuals who are deemed to have a higher rank or social standing in a pecking order based on a mutually valued set of social attributes. Higher social status or rank grants its holder a host of tangible benefits in both professional and personal domains. For instance, high-status actors are sought by groups for advice, are paid higher, receive unsolicited help, and are credited disproportionately in joint tasks. In innumerable ways, our social ecosystem consistently rewards those with high status.

A variety of elements contribute to status of potential offenders. In the academic world, business executives rise in status when they are invited as guest lecturers and keynote speakers at universities and conferences. Elite members rise in status when they have prestigious board positions at major corporations, and they rise in status when they serve on government committees. Furthermore, networking, friendships, and family ties can help in status achievement in many cultures. Philantrophy can also help improve status by donating money to charitable causes without claiming favors in return. A convicted white-collar offender in Norway who had received a medal from the King of Norway enjoyed such a high status that it was difficult for investigative journalists to work on the case, and it was difficult for the police to investigate and prosecute the criminal. This is in line with the classic article by Pontell et al. (2014) entitled "too big to fail, too powerful to jail".

Access includes legitimate access to strategic resources such as architects and attorneys (Adler & Kwon, 2002), specialized access in routine activity (Cohen & Felson, 1979), opportunity creation by entrepreneurship (Ramoglou & Tsang, 2016), and legitimate access to premises and systems (Benson & Simpson, 2018).

The municipality of Tjøme created opportunities by institutional deterioration (Rodriguez et al., 2005), inability to control because of social disorganization (Hoffmann, 2002), interference and noise in crime signals (Karim & Siegel, 1998), misrepresentation in documents (Qiu & Slezak, 2019), and failing control functions (Hurley et al., 2019). Chaos followed decay by lack of control in principal-agent relationships between politicians and bureaucrats (Bosse & Phillips, 2016), sense-making of actions was difficult for outsiders (Weick, 1995), costs exceeded benefits for potential whistleblowers in the municipality (Keil et al., 2010), and there was an ethical climate conflict (Victor & Cullen, 1988). Collapse resulted from rule complexity preventing compliance (Lehman et al., 2019), participation in networks by changing loyalties (Nielsen, 2003), and incremental applications and other deviant procedures became the usual way of business (Chang et al., 2005).

Convenient Personal Willingness

Offenders can conveniently justify crime and neutralize guilt feelings (Dearden, 2016, 2017, 2019). Application of neutralization techniques (Sykes & Matza, 1957), sliding on the slippery slope (Welsh et al. 2014), lack of self-control (Gottfredson & Hirschi, 1990), and narcissistic identification with the organization (Galvin et al., 2015) are some of the perspectives integrated in the willingness dimension of convenience theory. Learning from others by differential association (Sutherland, 1983) and professional deviant identity (Obodaru, 2017) are some further perspectives. When a white-collar offender justifies crime, then it is obvious to him and her that wrongdoing occurred. However, the offender can claim that the act of wrongdoing is morally justifiable (Schnatterly et al., 2018), and that a negative life event has occurred (Engdahl, 2015). When a white-collar offender denies a guilty mind, then the offender applies neutralization techniques. When a white-collar offender makes crime as a choice, it is convenient based on identity (Galvin et al., 2015), rationality (Pratt & Cullen, 2005), and learning from others (Sutherland, 1983). Social identity is an individual's self-concept as an organizational member (Piening et al., 2020).

In the case of illegal construction and expansion of summerhouses on the shoreline on Tjøme Island, including swimming pools, tennis courts, and parks, the willingness for environmental crime is a matter of choice and innocence as illustrated in Fig. 2.1. Offenders may find environmental crime convenient because it should be acceptable for individuals with elite identity (Petrocelli et al., 2003) combined with narcissistic traits expecting preferential treatment (Zvi & Elaad, 2018). An identity with reputation adaption to individual labels (Bernburg et al., 2006), with lack of social ties (Sampson & Laub, 1993), with deviant identity labeling (Mingus & Burchfield, 2012), with professional deviant identity (Obodaru, 2017), and narcissistic identification with the area (Galvin et al. 2015) can trigger willingness as well.

Offenders can make environmental crime a rational choice, where the perception of benefits exceeds the perception of costs (Pratt & Cullen, 2005), behavioral reinforcement of deviance develops over time (Benartzi et al., 2017), and undesirable impulses meet reluctance in self-regulation (Mawritz et al., 2017). Rationality can further derive from no perceived deterrence effect (Comey, 2009), stress that must be avoided at all costs (Olafsen et al., 2017), and sensation seeking to experience adventure (Craig & Piquero, 2017).

Offenders are learning from others by differential association (Sutherland, 1983), where the rich and mighty mansion owners on the island of Tjøme belong to a different vacation class from other people visiting the island during summer months. Deviant action might be according to authority as obedience (Baird & Zelin, 2009) where elite offenders have a collectivist value orientation (Bussmann et al., 2018).

Offenders may find their acts of wrongdoing morally justifiable (Schnatterly et al., 2018) based on upper echelon information selection (Gamache & McNamara, 2019). The feel entitled (Nichol, 2019), especially if negative life events have occurred (Engdahl, 2015) or if peer pressure exists (Gao & Zhang, 2019). Innocence from neutralization is possible by application of neutralization techniques (Sykes & Matza, 1957). Relevant neutralization arguments among environmental criminals on Tjøme Island include normality of action by claiming that everyone else does it, condemnation of the condemners, claim of entitlement, and claim of legal mistake. Some elite members may admit wrongdoing, but remove guilt feeling by claiming that they were sliding on the slippery slope (Welsh et al., 2014). Lack of self-control seems to be yet another factor that influences willingness for deviant behavior (Gottfredson & Hirschi, 1990).

Structural Convenience Model

Figure 2.1 illustrated the main themes in convenience theory. Figures 2.2, 2.3, 2.4, 2.5, and 2.6 illustrate several subthemes within each main theme of convenience theory.

Investigative Detection Journalism

Empirical research presented in this book draws on a number of primary and secondary sources. An important primary source for the environmental crime case in the municipality of Tjøme was two whistleblowers who supplied information for this research from 2017 to 2020. Whistleblowing is the disclosure by organizational members of illegal, immoral, or illegitimate practices under the control of their employers, to persons or organizations that may be able to effect action to terminate such practices (Wells et al., 2020). Whistleblowing is the disclosure by an individual in an organization or in society of deviant practices to someone who can do something about it (Andrade, 2015; Bjørkelo et al, 2011; Culiberg & Mihelic, 2017; Keil et al., 2010). Whistleblowing is an

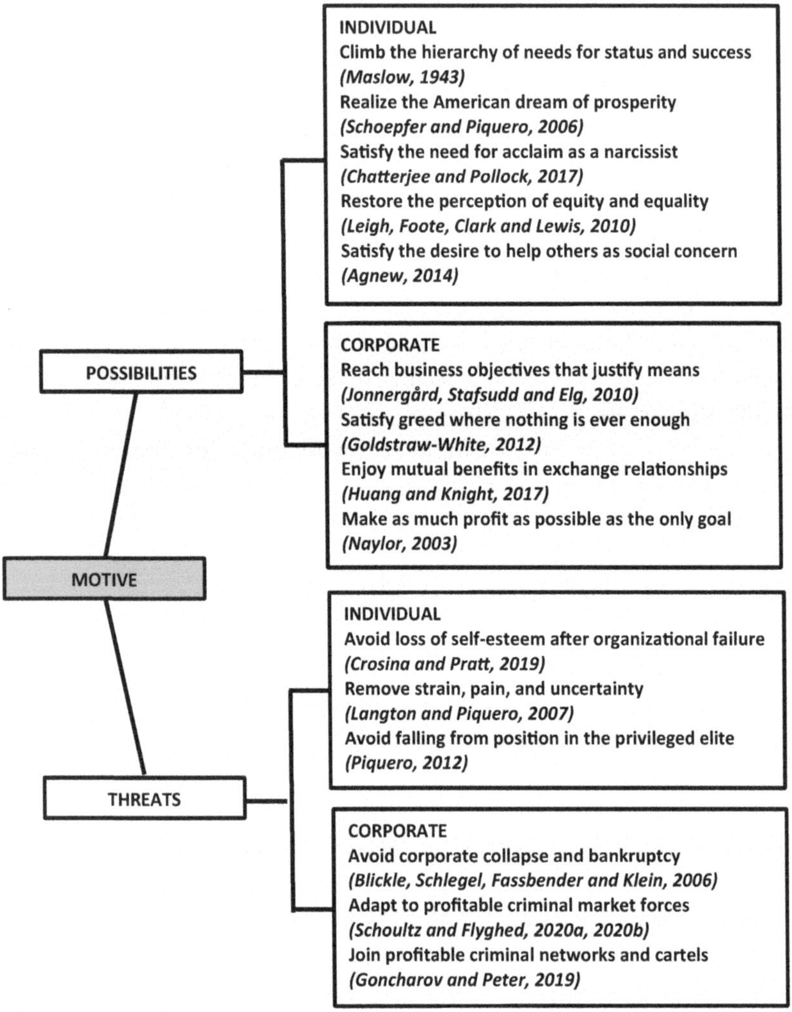

Fig. 2.2 Structural model of motive in convenience theory

action by employees who believe that their business or colleague(s) are involved in activities of misconduct and crime, cause unnecessary harm, transgress human rights, or contribute to otherwise immoral offenses (Mpho, 2017). However, one mechanism preventing many potential

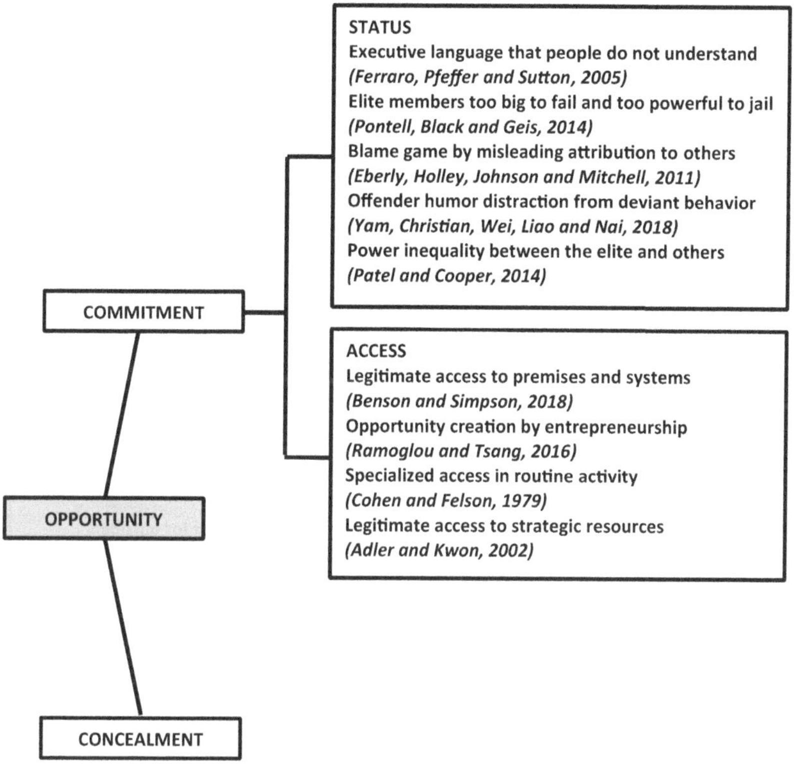

Fig. 2.3 Structural model of opportunity in convenience theory (part 1:2)

whistleblowers is the fear of retaliation and reprisals (Miceli & Near, 2013; Rehg et al., 2009).

An important secondary source was newspaper reports. Many Norwegian news media employ investigative journalists who report on misconduct and crime as cases evolve. Media coverage of the Tjøme case was extensive in the local newspaper *Tønsbergs Blad* and in the national business newspaper *Dagens Næringsliv*.

Empirical studies of media reports applied content analysis as the research method in the following chapters. Content analysis is any methodology or procedure that works to identify characteristics within texts attempting to make valid inferences (Krippendorff, 1980; Patrucco et al., 2017). Content analysis assumes that language reflects both how

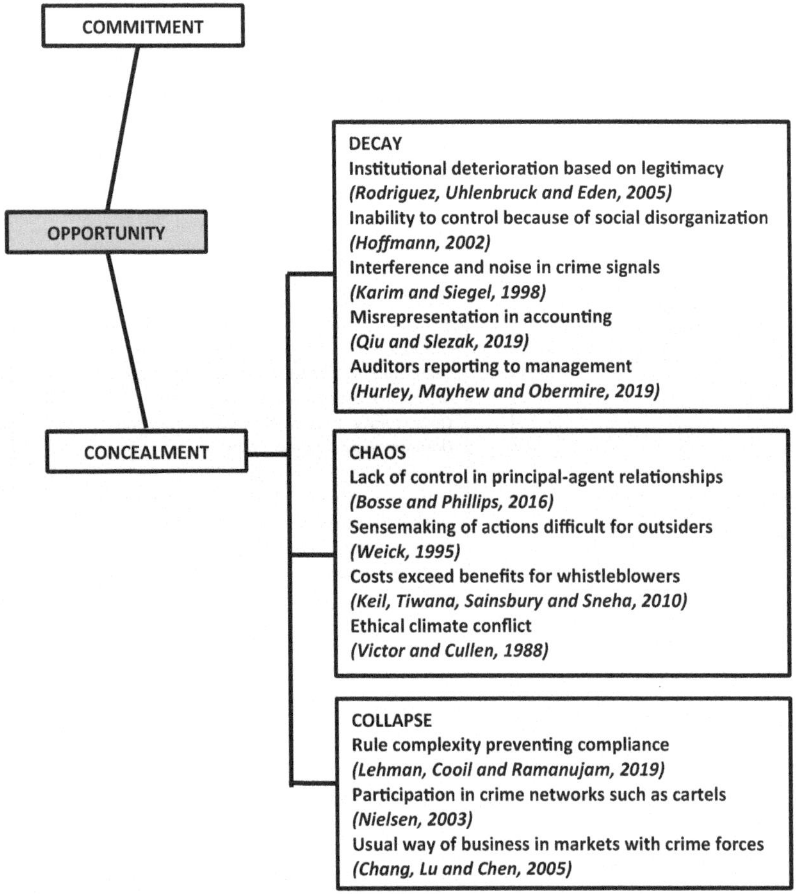

Fig. 2.4 Structural model of opportunity in convenience theory (part 2:2)

people understand their surroundings and their cognitive processes (Bell & Khoury, 2016; Saunders et al., 2007). Therefore, content analysis makes it possible to identify and determine relevant text in a context (McClelland et al., 2010).

As suggested by Barak (2007), news making criminology refers to the conscious efforts and activities of criminologists to make interpretations, to influence or to shape the representation of newsworthy items about crime and justice. News making criminology as a perspective on the

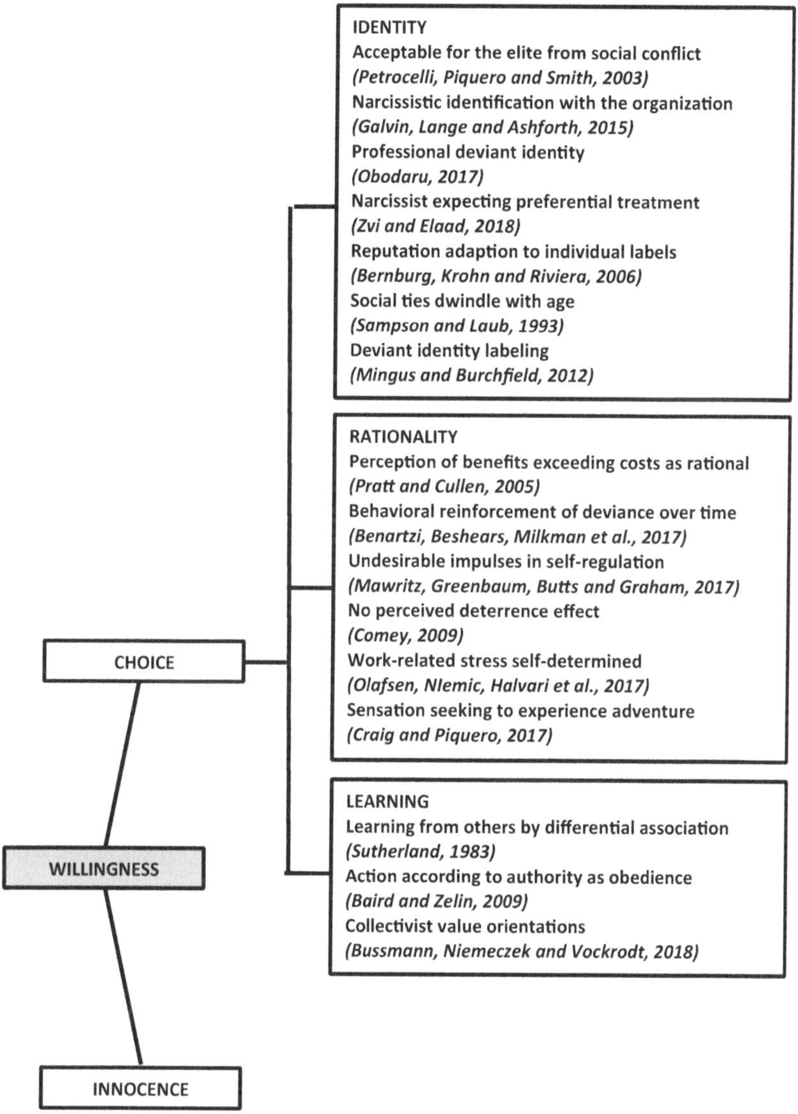

Fig. 2.5 Structural model of willingness in convenience theory (part 1:2)

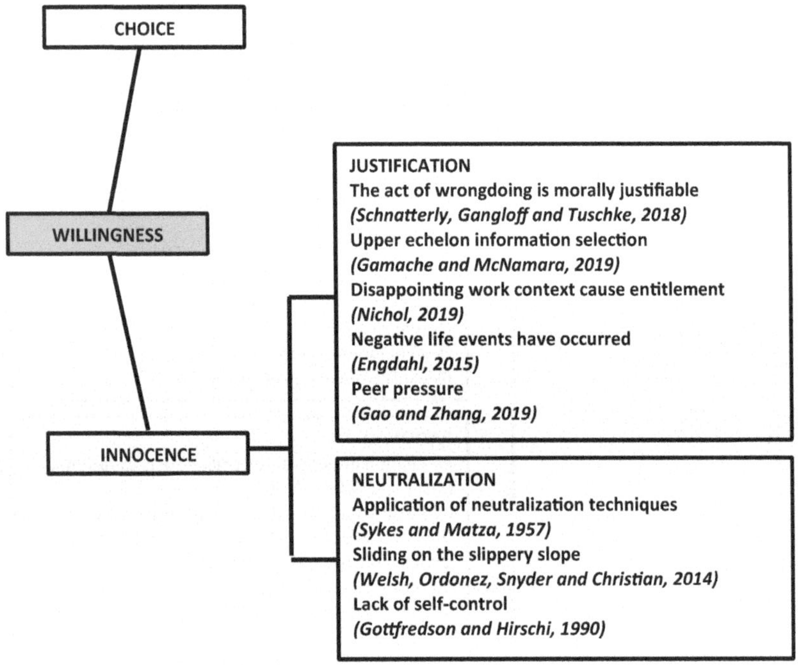

Fig. 2.6 Structural model of willingness in convenience theory (part 2:2)

theory, practice, and representations of crime and justice is an important approach for understanding white-collar crime. However, Barak's work focused on how the media constructs images of crime. This study uses the media as a source of potentially objective information, where facts in terms of quantitative numbers, names, and places from newspaper accounts become data for research. According to Burns and Orrick (2002), research suggests that the media present a distorted image of crime by focusing on violent, sensational events that are atypical of crime in society. They argue that the media is neglecting coverage of corporate offenses, and that the media disproportionately focus on conventional crime while neglecting the impact of corporate misbehavior. This line of reasoning does not only acknowledge possible biases in this research study, but it can also be understood as an argument for our research design, where an important characteristic of our sample is that the white-collar crime cases are prominent in the media.

Nevertheless, some types of corporate crime—probably those that are more typical—maybe still go unreported than other types of corporate crime. For instance, the media can suffer from bias against small corporate offenses preferring larger, more sensational offenses. Two methodological issues have to remain in mind because of our decision to use newspapers as sources:

- *Bias because of press coverage*. Financial crime committed by white-collar criminals is only interesting to the press to the extent that they are sensational and possibly revealed and discovered by the press itself. Therefore, no claim is justifiable that the sample is representative of white-collar criminals in general. Rather, there is a bias toward white-collar criminals that, for some reason, are of special interest to journalists and newspapers that cover their story. Therefore, the attribute of news coverage is explicitly part to the list of attributes for white-collar criminals including items such as position of trust and opportunity.
- *Data errors in press coverage*. Newspaper articles tend to have some errors in them. There may be factual errors, such as offender name, offender age, imprisonment sentence, crime type, and crime year. Furthermore, there may be a disproportionate focus on the sensational aspects of both criminal and crime. Anyone who has ever read in the newspaper about self will know that there are errors or at least wrong impressions created in the presentation. To minimize the source of error, several newspaper stories of the same case in different media were subject to review based on investigative research. Furthermore, we obtained court sentences in most of the cases to check both factual and story elements concerning both criminal and crime.

It is important to mention that journalists in Norway enjoy respectability because of their integrity and seriousness. Very few newspapers, if any, are engaged in reporting undocumented, sensational stories. In fact, during this research into financial crime by white-collar criminals, we have not found a single such newspaper in Norway. Some journalists in the financial press have developed sophisticated skills in digging for criminal cases, where they apply robust and transparent methodologies.

The media in other countries might find a very different public vision compared to the media in Norway, which enjoys and deserves public trust. For example, in the United States; "it now appears clear that some prominent columnists and commentators placed the trust at risk by accepting substantial fees from Enron" (Rosoff, 2009: 515). Furthermore, Knottnerus et al. (2006) argue that deviance at Enron could be difficult to detect because of special rituals that were an important part of corporate culture. The press, needless to say, depends on public trust (Pontell & Geis, 2007).

In Norway, the media coverage of white-collar crime in firms is typically according to the institutional perspective suggested by Graf-Vlachy et al. (2020: 43):

> Scholars in the tradition of institutional theory emphasize the role of the news media as an important part of firms' institutional environment. Scholars' underlying assumption here is that providing institutional and cultural accounts within which the appropriateness and desirability of actions can be evaluated … affects impression formation and the legitimation of firms. Moreover, these scholars assume that to arrive at their assessments of firms and firm actions, the media apply institutional logics held by their audiences and, thus, exert influential institutional pressures on firms.

Investigative journalists tend to develop hypotheses about phenomena and causality. They are very different from reporting journalists who only relate what they have heard or seen. Investigative journalists develop an idea via a study of potential offenders and their victims. They apply systematic analysis and treat their sources with care and professional concern.

In most criminal areas, people expect that a combination of victim and police are the main sources of criminal detection. After crime victims suffer an injury or a loss, they tend to report the incident to the police who investigate and potentially find the offender(s). In cases of financial crime by white-collar criminals, it is often quite different. A victim is frequently not aware of the injury or loss. For example, accounting fraud resulting in tax evasion is not a harm or damage perceived by tax authorities.

A number of angles find application in the process of white-collar crime detection within news media. On the one hand, we have the news media (newspapers and online-media) that have specialized and focused on financial information of all sorts, and report on this regularly. For them, the sources of information can be traditional through tips, company reports, stock-exchange information, and press conferences as well as other sources. For regular news media spread out over the country, the situation can be quite different. The detection of white-collar crime can come as a tip-off from a whistleblower or as official information if the police or an economic crime prosecutor performs a search locally. Whistleblowing is the disclosure by organizational members of illegal, immoral, or illegitimate practices under the control of their employers, to persons or organizations that may be able to effect action to terminate such practices (Wells et al., 2020). Whistleblowers in many cases alert journalists to serious crime and are sometimes the true detectors, not the journalists or media.

Additionally, the way the news develops into presentations in the media is dependent on many variables that occur at the same time: Do they have the right journalists in place at the time? Do they have an interest in the matter? Do they know anything or anyone related to this? A matter of resource balance typically takes place. The resource perspective in leading media houses is concerned with knowledge management.

Not many news media outside of the larger ones will have the possibility of setting aside journalists to work on an investigative white-collar crime for months. In the cases where they have done this, some experience among editors seems to be that there is an uncertainty as to whether this was worthwhile relative to the size and the complexity of the case. For a common, non-specialist news media, there will always be the balance of resources against the newsworthiness of the matter at hand. If a major white-collar crime story had emerged in Norway in the weeks after the July 22 terrorist attacks in 2011, reasonable doubt is justified whether the matter would have caught much attention in the general public press.

Most news media have a constant incoming flow of news on hand, and there is a constant daily priority of what is important and what deserves space in the publication. For all news items, there are some general rules of journalism that comes into play: Is it important for many people? Is it really news? Is it possible to get reliable information on this? Is it possible to approach the right people with the right questions? Can the investigative journalist approach both parties in a conflict? In addition to these questions, there will be a question as to whether the news organization at this specific point in time has the resources to deal with it. If the journalist knowledgeable of economic matters is on holiday, it is doubtful whether the news media organization will come back to the same matter later. That will depend on the development and the newsworthiness of the case at the second point in time. If the news organization is the first to report on the crime and it is regarded as "hot" it will probably do whatever possible to handle the matter at hand, knowing that other media, and especially online media can report on the same matter and as such "steal" the story. There is always an internal pride in a news organization when it can report on a matter of significant interest, and other news organizations cite their reports in the following news coverage.

The organizational culture also has an influence on white-collar crime detection among journalists. If you have journalists that find drive by their own interest to win investigative journalist prizes (SKUP in Norway), there is a higher possibility for such stories to emerge in publication. However, that will differ greatly among the news organizations. Øvrebø (2004) showed in a study of the Norwegian newspaper "Dagsavisen" after a change of editor-in-chief in 2001, that the news profile and priorities of the newspaper changed according to the principles laid down by the new editor when she took up her position. Personal preferences of an editor might have influence on the priorities of news in the newspaper, and that this will relate to all types of editorial material, whether it is general news, sports, culture, or financial news.

For a general news organization, white-collar crime is not a big story in itself unless it has repercussions on well-known persons locally or if something happens to the organization where the crime has taken place. Nationally it can be a big story if the person is a well-known profile, or if the crime in itself is of an unusual nature. If a main employer locally has to file for bankruptcy because of a white-collar crime, then the story is more than just another white-collar crime case since it has wider consequences that turn the world upside down for ordinary people in this local area. Then the white-collar crime will take the form of another typical important news story and be followed and treated as such, and the white-collar crime element ends up in a mix with other elements and consequential stories, building on the starting point as a white-collar crime. Campbell et al. (2001) studied the journalistic process of environmental news in Scotland, and addressed the information sources, which the news process explores and exploits. The study showed the preference for human sources as opposed to library-based information and discussed the influence of pragmatic constraints like time and space on the production of news. This process might likewise find application in the newsgathering process for white-collar crime.

The argument of white-collar crime detection among journalists seems to relate to the story's importance in itself and, as such, it might report as just another crime or news story and have the same internal process. For smaller news organizations without journalistic specialization in financial matters, the white-collar crime story might report according to the news prioritizing structure of that particular organization. For larger media organizations, which typically have separate sections for financial and economic news, the story becomes a matter within the prioritizing of that particular section. In addition, if the story is big enough in total, it will move from the particular section for finance into the general news of the organization. The higher the profile of persons involved, the more likely it is that it will have a more centralized coverage, i.e., moved into what is often the first section of the newspaper or the prioritized areas of a website's front page.

Four of the ten detector categories made up 62% of the total crime detecting sources and out of these the first two—journalists investigating tips from readers and crime victims suffering loss—made up 38% in Norway. This was the result of detection research, where 405 white-collar criminals ended up in prison from 2009 to 2015. Journalists detected one hundred and one offenders (25%). Not only was this the largest detector fraction, investigative journalist also detected more serious cases in terms of amount of money involved, compared to other detection categories such as bankruptcy lawyers and internal and external auditors.

It can be argued that the two categories of own investigations and victim stories are more susceptible to journalistic interest than the others, simply because it is easier to construct news stories based on these journalistic angles. Themes like manipulation and corruption are much more difficult to make into a story that is interesting for the readers simply because it is more complex and difficult to describe these matters in nonprofessional terms. A tip from readers that a news medium receives is most of the time accompanied by a subjective story from the person giving the tip that in turn gives the journalist clues to work on and discuss internally to assign the right news priority and angle. This argument finds support by the breakdown of 405 acts of white-collar crime showing that fraud is the category having the highest percentage of journalistic detection.

White-collar crime detection and follow-up seem to relate to a number of simultaneous journalistic procedures and cultural elements. For specialized publications in the financial information area, the white-collar crime news arena is closer at hand and the organization will typically be able to go deeper into the matter. If general or local news organizations detect white-collar crime, the procedure involved will more often take the form of a general news story with the resource balance that follows from that. It seems that journalists more often detect white-collar crime if the trigger is a tip from readers or if it is a case of fraud. Underlying all this are the internal news preferences and editorial guidance that is part of the policies of the news medium.

Finally, the most obvious reason for the high detection fraction by journalists is the fact that one of the criteria for the sample was newspaper coverage of the case. Naturally, this will lead to a bias toward journalist detection. However, in the stream of noise signals coming into newsrooms and editorial offices, it is impressive that investigative journalists are able to recognize the very few signals that have substance to be followed up by interviews and archival searches.

Bibliography

Adler, P. S., & Kwon, S. W. (2002). Social capital: Prospects for a new concept. *Academy of Management Review, 27*(1), 17–40.

Agnew, R. (2014). Social concern and crime: Moving beyond the assumption of simple self-interest. *Criminology, 52*(1), 1–32.

Andrade, J. (2015). Reconceptualising whistleblowing in a complex world. *Journal of Business Ethics, 128*(2), 321–335.

Baird, J. E., & Zelin, R. C. (2009). An examination of the impact of obedience pressure on perceptions of fraudulent acts and the likelihood of committing occupational fraud. *Journal of Forensic Studies in Accounting and Business, 1*(1), 1–14.

Barak, G. (2007). Doing newsmaking criminology from within the academy. *Theoretical Criminology, 11*(2), 191–207.

Bell, C. M., & Khoury, C. (2016). Organizational powerlessness, dehumanization, and gendered effects of procedural justice. *Journal of Managerial Psychology, 31*, 570–585.

Benartzi, S., Beshears, J., Milkman, K. L., Sunstein, C. R., Thaler, R. H., Shankar, M., Tucker-Ray, W., Congdon, W. J., & Galing, S. (2017). Should governments invest more in nudging? *Psychological Science, 28*(8), 1041–1055.

Benson, M. L., & Simpson, S. S. (2018). *White-collar crime: An opportunity perspective* (3rd ed.). Routledge.

Berdahl, J. L. (2007). Harassment based on sex: Protecting social status in the context of gender hierarchy. *Academy of Management Review, 32*(2), 641–658.

Bernburg, J. G., Krohn, M. D., & Rivera, C. (2006). Official labeling, criminal embeddedness, and subsequent delinquency. *Journal of Research in Crime and Delinquency, 43*(1), 67–88.

Bjørkelo, B., Einarsen, S., Nielsen, M. B., & Matthiesen, S. B. (2011). Silence is golden? Characteristics and experiences of self-reported whistleblowers. *European Journal of Work and Organizational Psychology, 20*(2), 206–238.

Blickle, G., Schlegel, A., Fassbender, P., & Klein, U. (2006). Some personality correlates of business white-collar crime. *Applied Psychology: An International Review, 55*(2), 220–233.

Bosse, D. A., & Phillips, R. A. (2016). Agency theory and bounded self-interest. *Academy of Management Review, 41*(2), 276–297.

Braaten, C. N., & Vaughn, M. S. (2019). Convenience theory of cryptocurrency crime: A content analysis of U.S. federal court decisions. *Deviant Behavior*, published online https://doi.org/10.1080/01639625.2019.1706706.

Burns, R. G., & Orrick, L. (2002). Assessing newspaper coverage of corporate violence: The dance hall fire in Qoteborg, Sweden. *Critical Criminology, 11*, 137–150.

Bussmann, K. D., Niemeczek, A., & Vockrodt, M. (2018). Company culture and prevention of corruption in Germany China and Russia. *European Journal of Criminology, 15*(3), 255–277.

Campbell, A., Muncer, S., & Bibel, D. (2001). Women and crime: An evolutionary approach. *Aggression and Violent Behavior, 6*(5), 481–497.

Chan, F., & Gibbs, C. (2020). Integrated theories of white-collar and corporate crime. In M. L. Rorie (Ed.), *The handbook of white-collar crime* (Chap. 13, pp. 191–208). Wiley.

Chang, J. J., Lu, H. C., & Chen, M. (2005). Organized crime or individual crime? Endogenous size of a criminal organization and the optimal law enforcement. *Economic Inquiry, 43*(3), 661–675.

Chatterjee, A., & Pollock, T. G. (2017). Master of puppets: How narcissistic CEOs construct their professional worlds. *Academy of Management Review, 42*(4), 703–725.

Clark, L. A., Foote, D. A., Clark, W. R., & Lewis, J. L. (2010). Equity sensitivity: A triadic measure and outcome/input perspectives. *Journal of Managerial Issues, 22*(3), 286–305.

Cohen, L. E., & Felson, M. (1979). Social change and crime rate trends: A routine activity approach. *American Sociological Review, 44*, 588–608.

Comey, J. B. (2009). Go directly to prison: White collar sentencing after the Sarbanes-Oxley Act. *Harvard Law Review, 122*, 1728–1749.

Craig, J. M. (2019). Extending situational action theory to white-collar crime. *Deviant Behavior, 40*(2), 171–186.

Craig, J. M., & Piquero, N. L. (2016). The effects of low self-control and desire-for-control on white-collar offending: A replication. *Deviant Behavior, 37*(11), 1308–1324.

Craig, J. M., & Piquero, N. L. (2017). Sensational offending: An application of sensation seeking to white-collar and conventional crimes. *Crime & Delinquency, 63*(11), 1363–1382.

Crosina, E., & Pratt, M. G. (2019). Toward a model of organizational mourning: The case of former Lehman Brothers bankers. *Academy of Management Journal, 62*(1), 66–98.

Culiberg, B., & Mihelic, K. K. (2017). The evolution of whistleblowing studies: A critical review and research agenda. *Journal of Business Ethics, 146*(4), 787–803.

Dearden, T. E. (2016). Trust: The unwritten cost of white-collar crime. *Journal of Financial Crime, 23*(1), 87–101.

Dearden, T. E. (2017). An assessment of adults' views on white-collar crime. *Journal of Financial Crime, 24*(2), 309–321.

Dearden, T. E. (2019). How modern psychology can help us understand white-collar criminals. *Journal of Financial Crime, 26*(1), 61–73.

Dearden, T. E., & Gottschalk, P. (2020). Gender and white-collar crime: Convenience in target selection. *Deviant Behavior*, published online https://doi.org/10.1080/01639625.2020.1756428.

Desai, N. (2020). Understanding the theoretical underpinnings of corporate fraud. *VIKALPA: The Journal for Decision Makers*, 1–7, published online https://doi.org/10.1177/0256090920917789,

Eberly, M. B., Holley, E. C., Johnson, M. D., & Mitchell, T. R. (2011). Beyond internal and external: A dyadic theory of relational attributions. *Academy of Management Review, 36*(4), 731–753.

Engdahl, O. (2015). White-collar crime and first-time adult-onset offending: Explorations in the concept of negative life events as turning points. *International Journal of Law, Crime and Justice, 43*(1), 1–16.

Farquhar, J. D., & Rowley, J. (2009). Convenience: A services perspective. *Marketing Theory, 9*(4), 425–438.

Ferraro, F., Pfeffer, J., & Sutton, R. I. (2005). Economics language and assumptions: How theories can become self-fulfilling. *Academy of Management Review, 30*(1), 8–24.

Friedrichs, D. O., Schoultz, I., & Jordanoska, A. (2018). *Edwin H. Sutherland, Routledge key thinkers in criminology*. Routledge.

Galvin, B. M., Lange, D., & Ashforth, B. E. (2015). Narcissistic organizational identification: Seeing oneself as central to the organization's identity. *Academy of Management Review, 40*(2), 163–181.

Gamache, D. L., & McNamara, G. (2019). Responding to bad press: How CEO temporal focus influences the sensitivity to negative media coverage of acquisitions. *Academy of Management Journal, 62*(3), 918–943.

Gao, P., & Zhang, G. (2019). Accounting manipulation, peer pressure, and internal control. *The Accounting Review, 94*(1), 127–151.

Goldstraw-White, J. (2012). *White-collar crime: Accounts of offending behavior*. Palgrave Macmillan.

Goncharov, I., & Peter, C. D. (2019). Does reporting transparency affect industry coordination? Evidence from the duration of international cartels. *The Accounting Review, 94*(3), 149–175.

Gottfredson, M. R., & Hirschi, T. (1990). *A general theory of crime*. Stanford University Press, CA.

Gottschalk, P. (2017). Convenience in white-collar crime: Introducing a core concept. *Deviant Behavior, 38*(11), 605–619.

Gottschalk, P. (2019). *Convenience triangle in white-collar crime—Case studies of fraud examinations*. Edward Elgar.

Graf-Vlachy, L., Oliver, A. G., Banfield, R., König, A., & Bundy, J. (2020). Media coverage of firms: Background, integration, and directions for research. *Journal of Management, 46*(1), 36–69.

Higgins, E. T. (1997). Beyond pleasure and pain. *American Psychologist, 52*, 1280–1300.

Hoffmann, J. P. (2002). A contextual analysis of differential association, social control, and strain theories of delinquency. *Social Forces, 81*(3), 753–785.

Holtfreter, K., Beaver, K. M., Reisig, M. D., & Pratt, T. C. (2010). Low self-control and fraud offending. *Journal of Financial Crime, 17*(3), 295–307.

Huang, L., & Knight, A. P. (2017). Resources and relationships in entrepreneurship: An exchange theory of the development and effects of the entrepreneur-investor relationship. *Academy of Management Review, 42*(1), 80–102.

Huisman, W., & Erp, J. (2013). Opportunities for environmental crime. *British Journal of Criminology, 53*, 1178–1200.

Hurley, P. J., Mayhew, B. W., & Obermire, K. M. (2019). Realigning auditors' accountability: Experimental evidence. *The Accounting Review, 94*(3), 233–250.

Jonnergård, K., Stafsudd, A., & Elg, U. (2010). Performance evaluations as gender barriers in professional organizations: A study of auditing firms. *Gender, Work and Organization, 17*(6), 721–747.

Jordanoska, A. (2018). The social ecology of white-collar crime: Applying situational action theory to white-collar offending. *Deviant Behavior, 39*(11), 1427–1449.

Kakkar, H., Sivanathan, N., & Globel, M. S. (2020). Fall from grace: The role of dominance and prestige in punishment of high-status actors. *Academy of Management Journal, 63*(2), 530–553.

Karim, K. E., & Siegel, P. H. (1998). A signal detection theory approach to analyzing the efficiency and effectiveness of auditing to detect management fraud. *Managerial Auditing Journal, 13*(6), 367–375.

Kawasaki, T. (2020). Review of comparative studies on white-collar and corporate crime. In M. L. Rorie (Ed.), *The handbook of white-collar crime* (Chap. 27, pp. 437–447). Wiley.
Keil, M., Tiwana, A., Sainsbury, R., & Sneha, S. (2010). Toward a theory of whistleblowing intentions: A benefit-cost differential perspective. *Decision Sciences, 41*(4), 787–812.
Klenowski, P. M., Copes, H., & Mullins, C. W. (2011). Gender, identity, and accounts: How white collar offenders do gender when making sense of their crimes. *Justice Quarterly, 28*(1), 46–69.
Knottnerus, J. D., Ulsperger, J. S., Cummins, S., & Osteen, E. (2006). Exposing Enron: Media representations of ritualized deviance in corporate culture. *Crime Media Culture, 2*(2), 177–195.
Krippendorff, K. (1980). *Content analysis: An introduction to its methodology.* Sage.
Langton, L., & Piquero, N. L. (2007). Can general strain theory explain white-collar crime? A preliminary investigation of the relationship between strain and select white-collar offenses. *Journal of Criminal Justice, 35*(1), 1–15.
Lehman, D. W., Cooil, B., & Ramanujam, R. (2019). The effects of rule complexity on organizational noncompliance and remediation: Evidence from restaurant health inspections. *Journal of Management*, 1–33, published online, https://doi.org/10.1177/0149206319842262.
Leigh, A. C., Foote, D. A., Clark, W. R., & Lewis, J. L. (2010). Equity sensitivity: A triadic measure and outcome/input perspectives. *Journal of Managerial Issues, 22*(3), 286–305.
Leonard, W. N., & Weber, M. G. (1970). Automakers and dealers: A study of criminogenic market forces. *Law & Society Review, 4*(3), 407–424.
Logan, M. W., Morgan, M. A., Benson, M. L., & Cullen, F. T. (2019). Coping with imprisonment: Testing the special sensitivity hypothesis for white-collar offenders. *Justice Quarterly, 36*(2), 225–254.
Mai, H. T. X., & Olsen, S. O. (2016). Consumer participation in self-production: The role of control mechanisms, convenience orientation, and moral obligation. *Journal of Marketing Theory and Practice, 24*(2), 209–223.
Maslow, A. H. (1943). A theory of human motivation. *Psychological Review, 50*(4), 370–396.
Mawritz, M. B., Greenbaum, R. L., Butts, M. M., & Graham, K. A. (2017). I just can't control myself: A self-regulation perspective on the abuse of deviant employees. *Academy of Management Journal, 60*(4), 1482–1503.
McClanahan, B., & South, N. (2020). 'All knowledge begins with the senses': Towards a sensory criminology. *British Journal of Criminology, 60*, 3–23.

McClean, E. J., Martin, S. R., Emich, K. J., & Woodruff, T. (2018). The social consequences of voice: An examination of voice type and gender on status and subsequent leader emergence. *Academy of Management Journal, 61*(5), 1869–1891.

McClelland, P. L., Liang, X., & Barker, V. L. (2010). CEO commitment to the status quo: Replication and extension using content analysis. *Journal of Management, 36*(5), 1251–1277.

Miceli, M. P., & Near, J. P. (2013). An international comparison of the incidence of public sector whistle-blowing and the prediction of retaliation: Australia Norway, and the US. *Australian Journal of Public Administration, 72*(4), 433–446.

Mingus, W., & Burchfield, K. B. (2012). From prison to integration: Applying modified labeling theory to sex offenders. *Criminal Justice Studies, 25*(1), 97–109.

Mpho, B. (2017). Whistleblowing: What do contemporary ethical theories say? *Studies in Business and Economics, 12*(1), 19–28.

Naylor, R. T. (2003). Towards a general theory of profit-driven crimes. *British Journal of Criminology, 43*, 81–101.

Nichol, J. E. (2019). The effects of contract framing on misconduct and entitlement. *The Accounting Review, 94*(3), 329–344.

Nielsen, R. P. (2003). Corruption networks and implications for ethical corruption reform. *Journal of Business Ethics, 42*(2), 125–149.

Obodaru, O. (2017). Forgone, but not forgotten: Toward a theory of forgone professional identities. *Academy of Management Journal, 60*(2), 523–553.

Olafsen, A. H., Niemiec, C. P., Halvari, H., Deci, E. L., & Williams, G. C. (2017). On the dark side of work: A longitudinal analysis using self-determination theory. *European Journal of Work and Organizational Psychology, 26*(2), 275–285.

Onna, J. H. R. (2020). From the avalance to the game: White-collar offenders on crime, bonds, and morality. *Crime, Law and Social Change*, published online https://doi.org/10.1007/s10611-020-09899-x.

Onna, J. H. R., & Denkers, A. J. M. (2019). Social bonds and white-collar crime: A two-study assessment of informal social controls in white-collar offenders. *Deviant Behavior, 40*(10), 1206–1225.

Øvrebø, T. (2004). Nyhetsproduksjon – kjønn og makt. En studie av endring i Dagsavisen 2000–2003 (News production—Sex and power. A study of change in Dagsavisen [*Norwegian Daily Newspaper*] 2000–2003). Hovedoppgave i Medievitenskap (Master thesis in Media Science). Universitetet i Oslo (University of Oslo).

Patel, P. C., & Cooper, D. (2014). Structural power equality between family and nonfamily TMT members and the performance of family firms. *Academy of Management Journal, 57*(6), 1624–1649.

Patrucco, A. S., Luzzini, D., & Ronchi, S. (2017). Research perspectives on public procurement: Content analysis of 14 years of publications in the Journal of Public Procurement. *Journal of Public Procurement, 16*(2), 229–269.

Petrocelli, M., Piquero, A. R., & Smith, M. R. (2003). Conflict theory and racial profiling: An empirical analysis of police traffic stop data. *Journal of Criminal Justice, 31*(1), 1–11.

Piening, E. P., Salge, T. O., Antons, D., & Kreiner, G. E. (2020). Standing together or falling apart? Understanding employees' responses to organizational identity threats. *Academy of Management Review, 45*(2), 325–351.

Piquero, N. L. (2012). The only thing we have to fear is fear itself: Investigating the relationship between fear of falling and white-collar crime. *Crime and Delinquency, 58*(3), 362–379.

Piquero, N. L. (2018). White-collar crime is crime: Victims hurt just the same. *Criminology & Public Policy, 17*(3), 595–600.

Pontell, H. N., & Geis, G. (2007). *International handbook of white-collar and corporate crime*. Springer.

Pontell, H. N., Black, W. K., & Geis, G. (2014). Too big to fail, too powerful to jail? On the absence of criminal prosecutions after the 2008 financial meltdown. *Crime, Law and Social Change, 61*(1), 1–13.

Pratt, T. C., & Cullen, F. T. (2005). Assessing macro-level predictors and theories of crime: A meta-analysis. *Crime and Justice, 32*, 373–450.

Qiu, B., & Slezak, S. L. (2019). The equilibrium relationships between performance-based pay, performance, and the commission and detection of fraudulent misreporting. *The Accounting Review, 94*(2), 325–356.

Ramoglou, S., & Tsang, E. W. K. (2016). A realist perspective of entrepreneurship: Opportunities as propensities. *Academy of Management Review, 41*, 410–434.

Rehg, M. T., Miceli, M. P., Near, J. P., & Scotter, J. R. V. (2009). Antecedents and outcomes of retaliation against whistleblowers: Gender differences and power relationships. *Organization Science, 19*(2), 221–240.

Rodriguez, P., Uhlenbruck, K., & Eden, L. (2005). Government corruption and the entry strategies of multinationals. *Academy of Management Review, 30*(2), 383–396.

Rosoff, S. M. (2009). The role of the mass media in the Enron fraud. In H. N. Pontell & G. Geis (Eds.), *International handbook of white-collar and corporate crime*. Springer.

Sampson, R. J., & Laub, J. H. (1993). *Crime in the making*. Harvard University Press.

Saunders, M., Lewis, P., & Thornhill, A. (2007). *Research methods for business students* (5th ed.). Pearson Education.

Schnatterly, K., Gangloff, K. A., & Tuschke, A. (2018). CEO wrongdoing: A review of pressure, opportunity, and rationalization. *Journal of Management, 44*(6), 2405–2432.

Schoepfer, A., & Piquero, N. L. (2006). Exploring white-collar crime and the American dream: A partial test of institutional anomie theory. *Journal of Criminal Justice, 34*(3), 227–235.

Schoultz, I., & Flyghed, J. (2020a). Denials and confessions: An analysis of the temporalization of neutralizations of corporate crime. *International Journal of Law, Crime and Justice*, published online https://doi.org/10.1016/j.ijlcj.2020.100389.

Schoultz, I., & Flyghed, J. (2020b). "We have been thrown under the bus": Corporate versus individual defense mechanisms against transnational corporate bribery charges, *Journal of White Collar and Corporate Crime*, published online https://doi.org/10.1177/2631309x20911883.

Sutherland, E. H. (1939). White-collar criminality. *American Sociological Review, 5*(1), 1–12.

Sutherland, E. H. (1983). *White collar crime—The uncut version*. Yale University Press.

Sykes, G. M., & Matza, D. (1957). Techniques of neutralization: A theory of delinquency. *American Sociological Review, 22*(6), 664–670.

Victor, B., & Cullen, J. B. (1988). The organizational bases of ethical work climates. *Administrative Science Quarterly, 33*, 101–125.

Weick, K. E. (1995). What theory is not, theorizing is. *Administrative Science Quarterly, 40*, 385–390.

Wells, J. B., Minor, K. I., Lambert, E. G., & Reeves, A. (2020). An exploratory study of possible correlates of individual whistleblowing propensity among sworn staff in a city jail. *Criminal Justice Policy Review*, 1–29, published online https://doi.org/10.1177/0887403420919478.

Welsh, D. T., Baer, M. D., Sessions, H.. & Garud, N. (2020). Motivated to disengage: The ethical consequences of goal commitment and moral disengagement in goal setting. *Journal of Organizational Behavior*, 1–15. https://doi.org/10.1002/job.2467.

Welsh, D. T., Ordonez, L. D., Snyder, D. G., & Christian, M. S. (2014). The slippery slope: How small ethical transgressions pave the way for larger future transgressions. *Journal of Applied Psychology, 100*(1), 114–127.

Wikstrom, P. O. H., Mann, R. P., & Hardie, B. (2018). Young people's differential vulnerability to criminogenic exposure: Bridging the gap between people- and place-oriented approaches in the study of crime causation. *European Journal of Criminology, 15*(1), 10–31.

Williams, M. L., Levi, M., Burnap, P., & Gundur, R. V. (2019). Under the corporate radar: Examining insider business cybercrime victimization through an application of routine activities theory. *Deviant Behavior, 40*(9), 1119–1131.

Yam, K. C., Christian, M. S., Wei, W., Liao, Z., & Nai, J. (2018). The mixed blessing of leader sense of humor: Examining costs and benefits. *Academy of Management Journal, 61*(1), 348–369.

Yeok, S. G., Ngah, K., Mustapha, J., & Hamid, K. A. (2020). Assessing the efficacy of rehabilitation programs for white-collar criminals: A case study of the northern states of Malaysia. *International Journal of Social Science Research, 2*(1), 109–123.

Zvi, L., & Elaad, E. (2018). Correlates of narcissism, self-reported lies, and self-assessed abilities to tell and detect lies, tell truths, and believe others. *Journal of Investigative Psychology and Offender Profiling, 15*, 271–286.

3
Vest Tank Case Study

This chapter tests the theory of convenience by a case study of three convicted individuals after treatment of desulphurization waste. Both the chief executive officer and the chairperson of the board at Vest Tank in Norway, as well as a chemical advisor to the company, received prison sentences for the case presented in this chapter (Gormley, 2009; Gulating, 2013, 2015; Nordhordland, 2010; Pedersen, 2017). The verdict is an exception rather than the rule when it comes to illegitimate financial savings and rewards from environmental crime by white-collar offenders (Böhm, 2020; Lynch, 2020). For example, after the Deepwater Horizon oil spill in the Gulf of Mexico in 2010, prosecutors brought criminal charges against four British Petroleum executives, but no one ended up in prison (Fowler, 2014; Freeh, 2013; Kang & Thosuwanchot, 2017; Thompson, 2017).

This chapter applies the theory of convenience to the case of Vest Tank pollution in Norway. Convenience theory is an emerging perspective on white-collar crime that provides insights into motive, opportunity, and willingness. This research is important as it applies the offender-based

P. Gottschalk, *Detecting and Investigating Environmental Crime*,
https://doi.org/10.1007/978-3-030-74184-6_3

perspective to environmental crime rather than the traditional offense-based perspective concerned with damage and harm (McClanahan & South, 2020).

Toxic Waste on Trafigura Ship

Parts of a tank installation operated by the company Vest Tank exploded in Sløvåg in Norway in 2007. The explosions caused illegal discharges that were injurious to health. The explosion generated an intense stench (NRK, 2015). Medical reactions developed and subsequent health complaints arrived from approximately 200 local residents around Sløvåg north of the city of Bergen. The pollution control authority ('Miljødirektoratet') reported Vest Tank to the prosecution authority for environmental crime ('Økokrim'). Miljødirektoratet stated that the company did not have a permit to perform desulphurization of gasoline, and the explosion occurred because of the activity to process the waste (Bakke & Knudssøn, 2008, 2009). Miljødirektoratet stated that Vest Tank had acted with gross negligence (Gormley, 2009). Økokrim brought charges against the chairperson of the board and the general manager at Vest Tank for environmental crime, as well as a chemical advisor to Vest Tank for breach of the pollution control act in Norway (Pedersen, 2017).

Vest Tank was a company specialized in the receipt and treatment of wash water from the cleaning of empty tanks on oil tankers. The company held the required environmental permits for such activities. However, Vest Test started to perform desulphurization of gasoline in 2006, which resulted in large quantities of desulphurization waste. The attempt to treat this waste led to the explosion (Gormley, 2009; Leer-Salvesen & Knudssøn, 2008). The offshore oil trader Trafigura supplied the waste (Oliver, 2010), and Trafigura lawyers tried to stop the media from publishing the scandal stories of dumping practices about the company (Knudssøn & Bakke, 2009):

> Trafigura lawyers also targeted our storytelling in Norway. However, here press freedom is more robust. In the struggle to reveal details of the waste dumping in Ivory Coast, the Guardian and the BBC are not alone in

> attracting the attentions of Trafigura's lawyers brandishing gags. Here in Norway, at the national broadcaster, we too received stern legal warnings. Therefore, we published.
>
> Norway has an important, if little reported, role in the Trafigura scandal. Back in 2006, as the Probo Koala tanker ship delivered waste to Ivory Coast, her sister ship, Probo Emu, was preparing for the same journey. However, when controversy emerged in the West African state, Trafigura redirected the vessel to a tank facility in Norway.

While investigative journalists in the United Kingdom and the Netherlands traced the toxic waste to Trafigura (Knudssøn & Bakke, 2009; Oliver, 2010), the criminal justice system in Norway prosecuted responsible individuals locally in the country (Bakke & Knudssøn, 2008, 2009; Nordhordland, 2010). Trond Emblem, owner and chairperson of the board at Vest Tank, was in 2013 sentenced to 2 years and 5 months in prison for violations of the pollution control act, the working environment act, and the fire and explosion protection act (Gulating, 2013: 14):

> There is sufficient evidence that Emblem had relatively sufficient information about operations, including operational problems, in Vest Tank.

Jostein Berland, chief executive officer at Vest Tank, was in 2012 sentenced to 2 years in prison for violations of the pollution control act, the working environment act, and the fire and explosion protection act (Gulating, 2013: 4):

> As CEO of Vest Tank, he participated in the decision and implementation of Vest Tank receiving and purifying / desulphurization of coker gasoline, which is essentially low octane gasoline and high sulfur content.

Karl-Jan Erstad, advisor to Vest Tank with a doctoral degree in chemicals, was in 2013 sentenced to 90 days in jail for violation of the pollution control act (Gulating, 2013: 23):

> There is no doubt that Erstad has good theoretical knowledge in his field of study (…) When Erstad applied to be an advisor in connection with the waste project, this was in an unknown area for him.

Alexela, a company that took over Vest Tank the day before the explosion, accepted a fine of two million Norwegian kroner ($ 240.000). Alexela has its headquarters in Estonia and operates within fuel retailing.

Environmental Crime Convenience

Returning to the structural model of convenience theory in Fig. 2.1, this research suggests that some of the themes are relevant in the case, while other themes are not, as illustrated in Fig. 3.1.

The motive for Emblem and Berland was profit for both of them. They wanted to develop Vest Tank into a profitable business and then sell the business to Alexela with profit. While Berland would receive compensation after the sale as an individual, Emblem's parent company Vest Kran would make money from the sale, as concluded by the district court (Nordhordland, 2010: 73):

> The court can see no other motives for the business that Berland managed than a desire to make the business profitable and then sell Vest Tank at a good price. Berland had entered into an agreement with Emblem concerning percentages of profits from the sale.

Therefore, in Fig. 3.1, both individual and corporate possibilities are themes of motive relevant to the case of Vest Tank. The profit would benefit Berland as an individual and Emblem as a corporate owner.

In the organizational opportunity perspective to commit white-collar crime, Emblem and Berland had access to resources to commit crime. They bought tanks for waste disposal, and they accepted waste from Trafigura.

In the organizational opportunity perspective to conceal white-collar crime, lack of oversight and guardianship causes chaos. Lack of oversight and guardianship might result from agency problems suggesting that a

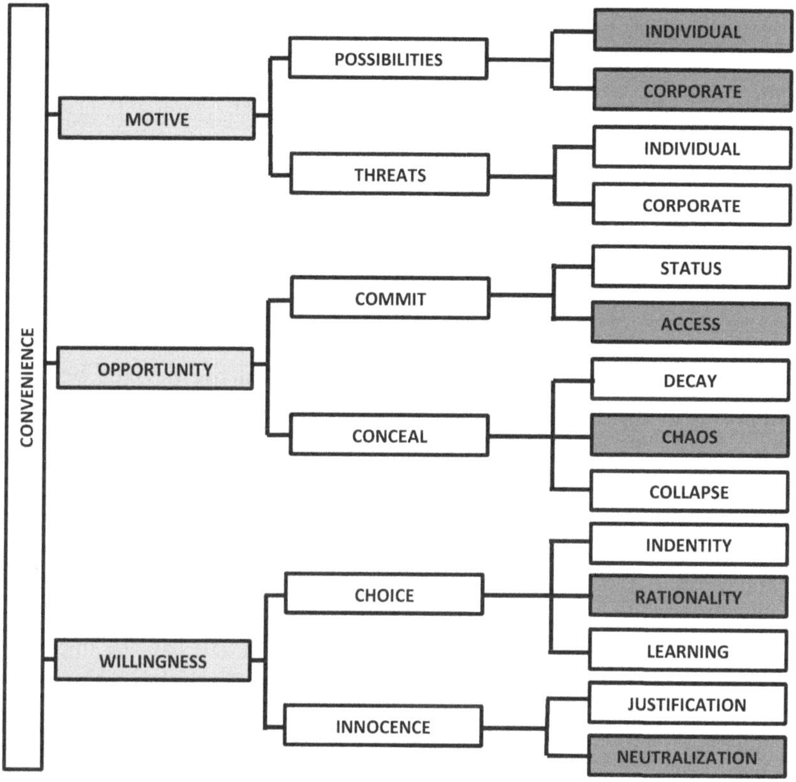

Fig. 3.1 Convenience themes in the case of Vest Tank

principal is often unable to control an agent who does work for the principal. The principal-agent model assumes narrow self-interest among both principals and agents. The interests of principal and agent tend to diverge, they may have different risk willingness or risk aversion, there is knowledge asymmetry between the two parties, and the principal has imperfect information about the agent's contribution (Bosse & Phillips, 2016; Chrisman et al., 2007; Pillay & Kluvers, 2014; Williams, 2008).

Emblem and Berland made Vest Tank attractive to the new owner Alexela. Alexela is thus the principal in the relationship, while Emblem and Berland are agents. As stated in court documents, Emblem performed a betrayal (Gulating, 2015: 9):

> Trond Emblem performed the betrayal and was at that time chairperson of Vest Tank. The starting point is that a chairperson bears identity with the company.

The court concluded that Emblem acted contrary to the interests of the company. Similarly, both Berland and Erstad acted contrary to the interests of Vest Tank because "the reception, cleaning and loading posed a specific danger of contamination due to the possibility of an explosion and / or fire in the plant" (Gulating, 2015: 10).

In the personal willingness perspective, statements made by the defendants in court are important to understand the deviant behavior (Gulating, 2013: 15):

> Emblem has explained that he was involved in the decision-making process, but then only in relation to the question of receiving waste would damage the company's reputation. He conferred with Berland on this, and knew that Berland had contacted pollution authorities about the reception. Berland has explained that he had contact with pollution authorities about what the company had authorization to do, and had the impression that the expansion of the business had received clearance. This was according to Berland announced to Emblem.

Emblem claimed lack of knowledge and lack of involvement (Nordhordland, 2010: 39):

> Emblem explained in court that he did not consider whether washing of coker gasoline was within the permits (…) Emblem has stated that he left the assessment to the general manager and that he relied fully on Berland (…) Emblem has explained that Berland was always very busy, and that he knew Berland as a "super optimist".

There is little evidence of crime as a choice based on identity or learning, as suggested in Fig. 3.1. The defendants in court claimed that they understood contacts with Norwegian authorities to imply approval for handling new kinds of gasoline waste. In addition, Trafigura had not told them the whole and true story about the characteristics of the coker gasoline and the risks involved in handling it. However, as pointed out in court documents, there was negligence on the part of all three defendants regarding obvious concerns, where they seemed reluctant to look into the matter beforehand. Their reluctant behavior might be interpreted as a rational choice, as indicated in Fig. 3.1. The rationality of considering pros and cons, benefits and costs (Pillay & Kluvers, 2014; Pratt & Cullen, 2005; Welsh et al., 2014), based on limited information and assumptions seems present in this case. There is more evidence of perceived innocence in the willingness dimension. The defendants applied a number of neutralization techniques (Jordanoska, 2018; Kaptein & Helvoort, 2019; Schoultz & Flyghed, 2020a, 2020b). First, they disclaimed responsibility for the crime. While Emblem claimed, he did not really know what was going on; Berland claimed that he thought he had approval, while Erstad claimed he was simply an advisor with no responsibility for actions by the company. The defendants thus claimed that one or more of the conditions of responsible agency did not occur.

Second, both Berland and Erstad claimed higher loyalties according to expectations. Those who feel they are in a dilemma employ this technique to indicate that the dilemma must be resolved at the cost of violating a law or policy. Emblem expected from Berland and Erstad contributions that made Alexela pay a high price for the procurement of Vest Tank.

Third, all three defendants claimed entitlement to action based on their insights and the perceived approval from pollution authorities. Waste disposal is a business with certain risks, and Vest Tank was a certified company for this kind of business.

Fourth and final, the defendants perceived being victims, rather than offenders. They claimed being victims of the incident that ruined their lives. The incident led to police investigation, prosecution, and incarceration. The media was publishing pictures of the defendants and condemning their actions in advance of court hearings. The defendants perceived being losers and being transformed into victims of those who reacted to their actions after the explosion.

Deductive Convenience Integration

So far, the analysis of Emblem says that he was greedy, motivated by profit and had lack of oversight from the new owner. The complicity of other actors, such as Berland and Erstad, contributes to the larger picture of Emblem's misconduct and crime.

The theory of convenience is an emerging approach to cross-level deductive integration of numerous perspectives on white-collar crime (Chan & Gibbs, 2020). It is a combination of many preexisting perspectives, which are relevant for selection based on their perceived commonalities. To integrate is to form, coordinate, or blend into a functioning or unified whole. Integration is to add perspectives and propositions that improve the validity, generalizability, and utility of a theory to explain a phenomenon and to predict potential outcomes (Fried & Slowik, 2004; Hambrick & Lovelace, 2018).

The theory of convenience integrates various perspectives on convenience into a single theory with greater comprehensiveness and explanatory value than any one of its component perspectives. As such, convenience theory attempts to explain white-collar crime by bringing together several different theories and invoking multiple levels of analysis at the individual, organizational, as well as societal levels as suggested by Friedrichs (2010: 479):

> The number of different theories or levels, and the formality, with which the relationship between the theories or variables on different levels of analysis is posited, varies.

Deductive reasoning is the process of reasoning from one or more statements (premises) to reach a logically certain conclusion. A conclusion results reductively by applying general rules, which hold over the entirety of a closed domain of discourse. Pratt et al. (2019) argue that most management scholars employ deductive and quantitative methods in positivistic testing of hypotheses.

Krohn and Eassey (2014) argue that up-and-down integration—also referred to as deductive integration—might be considered the classic form of theoretical integration (Liska et al., 1989: 5):

> Up-and-down integration refers to identifying a level of abstraction or generality that encompasses much of the conceptualization of the constituent theories.

There are two types of up-and-down integration. In one type, integrated propositions in an up-and-down manner use the premise of one theory to derive the propositions of constituent theories. In particular, if theory A contains a more abstract or more general proposition from which theory B can be specified, then it is possible to integrate B with theory A, resulting in a single theory. Alternatively, if both theories A and B are able to be elaborated using assumptions more general than either theory individually, then it could be possible to synthesize the theories together in order to create a new theory C, which contains parts of both constituent theories. The former approach often involves theoretical reduction, while the latter involves theoretical synthesis.

The theory of convenience is a result of deductive integration by theoretical synthesis where a large number of theoretical perspectives blend into a unified whole to explain the main dimensions of motive, opportunity, and willingness. Some scholars argue that deductive integration is the only true type of theoretical integration. It is also the most difficult to do (Krohn & Eassey, 2014).

The theory of convenience also applies another kind of up-and-down integration by including different units of analysis. Some perspectives are at the individual level of the offender (e.g., personal neutralization and personal strain). Some perspectives are at the group level of the organization (e.g., lack of guardianship and institutional deterioration), while some perspectives are at the level of society (e.g., criminogenic market structures and resource scarcity).

In conclusion, the theory of convenience is an emerging approach to explain the occurrence of white-collar crime. This chapter tested the theory by a case study of environmental crime in Norway. Among the many themes included in the theory of convenience, the case study illustrated that some themes are relevant for the case while other themes are not. It was both occupational crime to benefit offenders and corporate crime to benefit Vest Tank. Offenders had access to resources to commit crime in terms of illegal waste disposal that caused an explosion harming many people. The illegal waste disposal was possible to conceal since the new owner was lacking oversight and guardianship. The offenders made a rational choice to take care of the dangerous waste. They felt innocent by disclaiming responsibility, claiming higher loyalty, claiming entitlement to action, and perceiving being victims rather than offenders.

There are several avenues for future research. First, it would be interesting to study the extent of convenience orientation among convicted white-collar offenders compared to other members of the elite in society. Next, it would be interesting to test the relative importance of the corners in the convenience triangle in terms of motive, opportunity, and willingness. For example, the extent of opportunity might be the major factor whether white-collar crime actually occurs or not.

Trafigura Toxic Pollution Crime

The government of the Ivory Coast took Trafigura to court, where the company was prosecuted and charged for 16 lethal cases and over 100,000 claims of health problems as a result of harm caused by toxic pollution (Maksimentsev & Maksimentseva, 2020: 286):

> In order to bring the story to a close and release its executives detained in a local prison in expectation of host state court rulings on criminal charges, Trafigura concluded a settlement deal, totaling 100b local francs, with host state government and local victims to pay 95b francs to civil victims and 5b francs to the Côted d'Ivoire state budget in reparation and compensatory payments, at that time an equivalent to approximately USD 198m; it also released its two top managers from the local jail after the deal was properly enacted on 12 February 2007 and countersigned by the government of Côte d'Ivoire. Also, in order to avoid proactive collective lawsuits from the extraterritorial legal attempts of 1,000 victims from the Abidjan community brought overseas to the London courts, Trafigura paid GBP 32m in an out-of-court settlement.

Courts not only in the Ivory Coast and the United Kingdom, but also in the Netherlands considered Trafigura as liable for deeply negligent pollution and damage to the safety of the living environment of the host state citizens. The courts found the damage to be a breach of security of industrial operations and transportation of oil products with a heavy impact on human health in Africa. Contact with toxic waste, spills, and sludge remaining in the soil and water along oil-transportation pipelines and around onshore and offshore oil-processing sites were considered the responsibility of both Royal Dutch Shell and Trafigura (Maksimentsev & Maksimentseva, 2020).

Rather than terminating their criminal activities after reactions and controversy in the West African state, Trafigura instead redirected the Probo Koala tanker ship to the tank facility Vest Tank in Norway (Knudssøn & Bakke, 2009; NRK, 2015; Oliver, 2010), as described above. By not telling the whole story about the cargo to executives at Vest Tank in Norway, the waste was accepted and treated by desulphurization (Gormley, 2009). Three Norwegians were incarcerated after the explosion at Vest Tank (Gulating, 2013, 2015; Nordhordland, 2010).

Maksimentsev and Maksimentseva (2020: 285) suggest that executives at Trafigura made a rational judgment to avoid risks of liability in the future by transferring responsibility to local actors in Africa:

> The rational judgment from the headquarters was to thus avoid risks related to physical participation in local subsoil use and the environmental impact of criminal extracting and mining operations. This would leave any trouble from host state jurisdictions to locally incorporated special-purpose vehicles (subsidiaries and affiliates), thus anticipating that the corruption and low competencies of local (host state) public officers, prosecution, courts and enforcement agents would limit or mitigate any potential risk and negative impact of environmental crimes of transnational corporations, with minimum or no material impact on mother companies, allowing them thus, according to Riley, to 'evade the risk of liability'.

Trafigura obviously considered it a rational choice to have operations through host state affiliates and subsidiaries under transnational companies holding corporate control.

Escalation of Goal Commitment

The actions of Trafigura management can be studied further by the concepts of goals and escalation of commitment to goals. Goalsetting is often perceived in a positive light, meaning that ambitious goals increase performance (Locke & Latham, 2013). However, there is also evidence suggesting that extreme goals can lead to unethical behavior (Jonnergård et al., 2010; Schweitzer et al., 2004; Simmons, 2018). Welsh et al. (2020) argue that this is not only because of rewards associated with goal achievement, but also because of changing moral reasoning processes related to goals over time. As such, strong goal commitment facilitates unethical behavior by increasing not only the motivation to achieve the goal, but also the motivation to justify doing so by any means necessary (Locatelli et al., 2017), where ends can justify means (Jonnergård et al., 2010; Welsh & Ordonez, 2014) For example, Dodge (2009: 15) suggests that tough rivalry among executives make them commit crime to attain goals: "The competitive environment generates pressures on the organization to violate the law in order to attain goals".

The ends-justify-means perspective is known as state morale disagreement, a process through which individuals justify unethical behavior (Moore, 2015). It is part of the dark side of ambitious goals. Escalation of commitment to ambitious goals increases the likelihood of misconduct and crime (Sleesman et al., 2012, 2018). White-collar offenders typically choose the most convenient path to reach their goals (Wikstrom et al., 2018).

Escalation of commitment to goals can be caused by a need to justify prior behavior, where a decision-maker may increase own commitment in the face of negative consequences from previous decisions. This higher level of commitment may in turn lead to further negative consequences. Thus, instead of accepting an immediate loss and termination of waste disposal as planned, decision-makers at Trafigura may have become desperate to complete the illegal waste disposal somehow and somewhere.

Biyalogorsky et al. (2006) argue that escalation is caused by the improper use of initial positive beliefs in the face of negative new information. Keil et al. (2010) noticed that negative status information is sometimes not available or not attended to. Simester and Zhang (2010) describe three ways of responding to unfavorable information in escalation settings; distorting, discrediting, or simply not collecting information that may reveal unfavorable news. Uncertainty, ambiguity, or simply lack of information about activity status may increase escalation. Sunk costs, self-efficacy, ego threats, and personal responsibility for the initial decision can all cause escalation (Sleesman et al., 2012).

To escalate is to increase, enlarge, or intensify an effort to reach a goal. In the Trafigura case of involving Vest Tank getting rid of the dangerous waste escalated to unethical and illegal practice.

Bibliography

Bakke, S., & Knudssøn, K. (2008, June 24). Coker gasoline—Low quality. Norwegian public broadcasting *NRK*. www.nrk.no.

Bakke, S., & Knudssøn, K. (2009, October 13). Trafigura and the Minton report. Norwegian public broadcasting *NRK*. www.nrk.no.

Biyalogorsky, E., Boulding, W., & Staelin, R. (2006). Stuck in the past: Why managers persist with new product failures. *Journal of Marketing, 70*(2), 108–121.

Böhm, M. L. (2020). Criminal business relationships between commodity regions and industrialized countries: The hard road from raw material to new technology. *Journal of White Collar and Corporate Crime, 1*(1), 34–49.

Bosse, D. A., & Phillips, R. A. (2016). Agency theory and bounded self-interest. *Academy of Management Review, 41*(2), 276–297.

Chan, F., & Gibbs, C. (2020). Integrated theories of white-collar and corporate crime. In M. L. Rorie (Ed.), *The handbook of white-collar crime* (Chap. 13, pp. 191–208). Wiley.

Chrisman, J. J., Chua, J. H., Breton-Miller, I., Miller, D., & Steier, L. (2018). Governance mechanisms and family firms. *Entrepreneurship Theory and Practice, 42*(2), 171–186.

Chrisman, J. J., Chua, J. H., Kellermanns, F. W., & Chang, E. P. C. (2007). Are family managers agents or stewards? An exploratory study in privately held family firms. *Journal of Business Research, 60*(10), 1030–1038.
Dodge, M. (2009). *Women and white-collar crime*. Prentice Hall.
Fowler, T. (2014, January 27). BP's new tactic in oil spill claims: Go after the 'special master'. *The Wall Street Journal*. www.wsj.com.
Freeh, L. J. (2013, September 6). *Independent external investigation of the Deepwater Horizon court supervised settlement program* (Report of Special Master Louis J. Freeh). www.laed.uscourts.gov.
Fried, Y., & Slowik, L. H. (2004). Enriching goal-setting theory with time: An integrated approach. *Academy of Management Review, 29*(3), 404–422.
Friedrichs, D. O. (2010). Integrated theories of white-collar crime. In F. T. Cullen & P. Wilcox (Eds.), *Encyclopedia of criminological theory* (Vol. 1, pp. 479–486). Sage.
Gormley, T. P. (2009, September 14). *Administrative and criminal penalties for illegal pollution under Pollution Control Act*. International Law Office. https://www.internationallawoffice.com/Newsletters/Environment-Climate-Change/Norway/Arntzen-de-Besche-Advokatfirma-AS/Administrative-and-Criminal-Penalties-for-Illegal-Pollution-under-Pollution-Control-Act.
Gulating. (2013, March 8). Court case LG-2011-60302—LG-2012-23847. *Gulating lagmannsrett* (Gulating court of appeals), Bergen, Norway.
Gulating. (2015, February 16). Court case 14-039291ASD-GULA/AVD1. *Gulating lagmannsrett (Gulating court of appeals)*, Bergen, Norway.
Hambrick, D. C., & Lovelace, J. B. (2018). The role of executive symbolism in advancing new strategic themes in organization: A social influence perspective. *Academy of Management Review, 43*(1), 110–131.
Jonnergård, K., Stafsudd, A., & Elg, U. (2010). Performance evaluations as gender barriers in professional organizations: A study of auditing firms. *Gender, Work and Organization, 17*(6), 721–747.
Jordanoska, A. (2018). The social ecology of white-collar crime: Applying situational action theory to white-collar offending. *Deviant Behavior, 39*(11), 1427–1449.
Kang, E., & Thosuwanchot, N. (2017). An application of Durkheim's four categories of suicide to organizational crimes. *Deviant Behavior, 38*(5), 493–513.
Kaptein, M., & Helvoort, M. (2019). A model of neutralization techniques. *Deviant Behavior, 40*(10), 1260–1285.

Keil, M., Tiwana, A., Sainsbury, R., & Sneha, S. (2010). Toward a theory of whistleblowing intentions: A benefit-cost differential perspective. *Decision Sciences, 41*(4), 787–812.

Knudssøn, K., & Bakke, S. (2009, October 26). A spreading of toxic gags. *The Guardian.* www.theguardian.com.

Krohn, M. D., & Eassey, J. M. (2014). Integrated theories of crime. In J. M. Miller (Ed.), *The encyclopedia of theoretical criminology* (pp. 458–463). Wiley.

Leer-Salvesen, T., & Knudssøn, K. (2008, June 24). A small pawn in the game. Norwegian public broadcasting *NRK*. www.nrk.no.

Liska, A. E., Krohn, M. D., & Messner, S. F. (1989). Strategies and requisites for theoretical integration in the study of crime and deviance. In S. F. Messner, M. D. Krohn, & A. E. Liska (Eds.), *Theoretical integration in the study of deviance and crime* (pp. 1–20). State University of New York Press.

Locatelli, G., Mariani, G., Sainati, T., & Greco, M. (2017). Corruption in public projects and megaprojects: There is an elephant in the room! *International Journal of Project Management, 35*(3), 252–268.

Locke, E. A., & Latham, G. P. (2013). Goal setting theory: The current state. In E. A. Locke & G. P. Latham (Eds.), *New developments in goal setting and task performance* (pp. 623–630). Routledge.

Lynch, M. J. (2020). Green criminology and environmental crime: Criminology that matters in the age of global ecological collapse. *Journal of White Collar and Corporate Crime, 1*(1), 50–61.

Maksimentsev, M., & Maksimentseva, N. (2020). The global impact of exercising extraterritorial jurisdiction over transnational corporate environmental crimes in extractive industries. *Law of Ukraine: Legal Journal, 1–2020*, 278–295.

McClanahan, B., & South, N. (2020). 'All knowledge begins with the senses': Towards a sensory criminology. *British Journal of Criminology, 60*, 3–23.

Moore, C. (2015). Moral disengagement. *Current Opinion in Psychology, 6*, 199–204.

Nordhordland. (2010, March 26). Court case 09-098460MED-NOHO. *Nordhordland tingrett (Nordhordland district court)*, Bergen, Norway.

NRK (2015, December 9). Dirty cargo. Norwegian public broadcasting *NRK*. https://tv.nrk.no/OAUA11005108.

Oliver, L. (2010, April 26). Trafigura investigators honored with Daniel Pearl Award. *Journalism.* www.journalism.co.uk.

Pedersen, N. (2017, August 19). "Professoren" sin kamp for å reinvaska seg ("The professor's" struggle to clean himself). Norwegian public broadcasting *NRK*. www.nrk.no.
Pillay, S., & Kluvers, R. (2014). An institutional theory perspective on corruption: The case of a developing democracy. *Financial Accountability & Management, 30*(1), 95–119.
Pratt, M. G., Kaplan, S., & Whittington, R. (2019). Editorial essay: The tumult over transparency: Decoupling transparency from replication in establishing trustworthy qualitative research. *Administrative Science Quarterly*, 1–19. https://doi.org/10.1177/0001839219887663.
Pratt, T. C., & Cullen, F. T. (2005). Assessing macro-level predictors and theories of crime: A meta-analysis. *Crime and Justice, 32*, 373–450.
Schoultz, I., & Flyghed, J. (2020a). Denials and confessions: An analysis of the temporalization of neutralizations of corporate crime. *International Journal of Law, Crime and Justice*, published online https://doi.org/10.1016/j.ijlcj.2020.100389.
Schoultz, I., & Flyghed, J. (2020b). "We have been thrown under the bus": Corporate versus individual defense mechanisms against transnational corporate bribery charges. *Journal of White Collar and Corporate Crime*, published online https://doi.org/10.1177/2631309x20911883.
Schweitzer, M. E., Ordóñez, L., & Douma, B. (2004). Goal setting as a motivator of unethical behavior. *Academy of Management Journal, 47*(3), 422–432.
Simester, D., & Zhang, J. (2010). Why are bad products so hard to kill? *Management Science, 56*(7), 1161–1179.
Simmons, A. (2018). Why students cheat and what to do about it. *Edutopia*. https://www.edutopia.org/article/why-students-cheat-andwhat-do-about-it.
Sleesman, D. J., Conlon, D. E., McNamara, G., & Miles, J. E. (2012). Cleaning up the big muddy: A meta-analytic review of the determinants of escalation of commitment. *Academy of Management Journal, 55*(3), 541–562.
Sleesman, D. J., Lennard, A. C., McNamara, G., & Conlon, D. E. (2018). Putting escalation of commitment in context: A multilevel review and analysis. *Academy of Management Annals, 12*(1), 178–207.
Thompson, R. (2017, October 29). With BP settlement claims winding down, Lafayette lawyer Parick Juneau turns attention to Takata air bag recall. *The Advocate*. www.theadvocate.com.

Welsh, D. T., & Ordonez, L. D. (2014). The dark side of consecutive high performance goals: Linking goal setting, depletion, and unethical behavior. *Organizational Behavior and Human Decision Processes, 123*, 79–89.
Welsh, D. T., Baer, M. D., Sessions, H., & Garud, N. (2020). Motivated to disengage: The ethical consequences of goal commitment and moral disengagement in goal setting. *Journal of Organizational Behavior*, 1–15. https://doi.org/10.1002/job.2467.
Welsh, D. T., Ordonez, L. D., Snyder, D. G., & Christian, M. S. (2014). The slippery slope: How small ethical transgressions pave the way for larger future transgressions. *Journal of Applied Psychology, 100*(1), 114–127.
Wikstrom, P. O. H., Mann, R. P., & Hardie, B. (2018). Young people's differential vulnerability to criminogenic exposure: Bridging the gap between people- and place-oriented approaches in the study of crime causation. *European Journal of Criminology, 15*(1), 10–31.
Williams, J. W. (2008). The lessons of Enron: Media accounts, corporate crimes, and financial markets. *Theoretical Criminology, 12*(4), 471–499.

4

Tjøme Island Case Study

This chapter applies the theory of convenience to white-collar offenses on the shoreline in Norway. The rich and mighty are occupying the most attractive recreational places along the coast with their vacation resorts in the form of summer palaces and private harbors. This chapter addresses the abuse of power associated with land development in an extremely attractive resort area in southern Norway. While Norwegian law makes it illegal to construct or expand buildings and other changes in the landscape in the one-hundred-meter belt along the coast, white-collar offenders hire architects, attorneys, and others to get what they want and defend what they have done. For the first time in recent history, a white-collar offender received a jail sentence for such environmental crime, which represents the case study presented in this chapter.

The shoreline is a limited public good that many nations protect and preserve for the recreation and benefit of all its inhabitants and visitors. Privatization of the shoreline, especially in the form of vacation resorts, is thus illegal. Nevertheless, some rich and mighty people are able to violate the law, and they build and expand their vacation homes on the shoreline. With the help of architects, builders, attorneys and relationships to community planners, they use their resources to get what they want.

P. Gottschalk, *Detecting and Investigating Environmental Crime*,
https://doi.org/10.1007/978-3-030-74184-6_4

This chapter presents a case study of a convicted white-collar offender in Norway (Agder, 2019; Berglihn, 2019). While the country has a very long coastline, the attractive spots for vacation houses and summer guests are located south of the capital of Oslo. In particular, the island of Tjøme is popular (BDO, 2017; DN, 2019). The theme of illegal construction and expansion of summerhouses serves as an example in relation to convenience theory, where the motive is to climb Maslow's hierarchy of needs, the opportunity is access to resources, and the willingness for deviant behavior is dependent on neutralization techniques, lack of self-control and other factors.

This chapter thus applies the theory of convenience to the case study. Convenience theory suggests that there is a motive, an opportunity, and a willingness among white-collar offenders to violate the law (Braaten & Vaughn, 2019; Chan & Gibbs, 2020; Dearden & Gottschalk, 2020; Gottschalk, 2017, 2019). The theory builds on the fraud triangle suggested by Cressey (1972). A white-collar offender is a member of the upper echelon in society who abuses the privileged position for illegitimate benefits (Sutherland, 1939, 1983).

Convenience theory is an emerging approach to integrate well-known perspectives on white-collar crime from criminology, sociology, and management. It is an integrated deductive theory (Chan & Gibbs, 2020; Friedrichs, 2010; Krohn & Eassey, 2014; Liska et al., 1989; Pratt et al., 2019). The choice of convenience theory in this chapter is not the result of a comparison to other perspectives such as power or class. Rather, as an emerging new theoretical perspective on white-collar offenders, it is interesting to explore its empirical application to a case of environmental crime.

This research is concerned with application of convenience theory to the case of rich and mighty white-collar offenders who commit environmental crime by privatizing and harming the shoreline in terms of environmental crime. Specifically, this research attempts to identify convenience themes that seem present in the case of one convicted summerhouse owner.

Two whistleblowers and some investigative journalists started to perceive crime signals. The theory of crime signals suggests that detection is dependent on signal intensity, signal alertness, pattern recognition, and personal knowledge (Karim & Siegel, 1998). The signal intensity from illegal construction sites was low, except when people could observe construction work from boats on the sea. The local population had declining signal alertness, as they knew that no corrections would occur anyway. The whistleblowers as informants and notifiers were able to recognize patterns and identify key persons who were involved in all recent cases. The persons worked in the municipality as planners and as private architects.

This archival research is based on content analysis of newspaper reports (e.g., Berglihn, 2019; DN, 2019; Holm, 2020; Holmøy, 2018; Klevstrand, 2018; Lien & Aunemo, 2019; Wang, 2019), municipal files (e.g., Tilsynsrapport, 2020), fraud examinations (e.g., BDO, 2017), and court documents (e.g., Agder, 2019). Content analysis is any methodology or procedure that works to identify characteristics within texts attempting to make valid inferences (Krippendorff, 1980; Patrucco et al., 2017). Content analysis assumes that language reflects both how people understand their surroundings and their cognitive processes (Bell & Khoury, 2016; Saunders et al., 2007). Therefore, content analysis makes it possible to identify and determine relevant text in a context (McClelland et al., 2010).

The Island of Tjøme

Our case is concerned with illegally constructed summerhouses and expanded vacation resorts on the beaches and cliffs in the southern part of Norway. Both environmental protection laws and public access laws have made it illegal to build and expand vacation homes in the hundred meters wide belt along the coast. Nevertheless, rich people have been able to ignore the building ban in the case municipality of Tjøme. The environmental harm from shoreline violations is twofold. First, public access to attractive recreational areas suffers harm by illegal privatization. Public health and democracy suffer harm. Second, construction work changes the landscape. Nature, animals, and wildlife thus suffer harm as well. The beach zone is important for the diversity of nature, for outdoor life, and the public's access to the waterfront, and it is of national interest that it be taken care of (Klevstrand & Solem, 2020). The impacts to the natural environment are often extensive and long lasting (Gibbs & Boratto, 2017).

The municipality of Tjøme is located on an island south of the Norwegian capital, Oslo. The island is the most expensive summerhouse resort and may deserve a comparison to Cape Cod in the United States in terms of its attractiveness for people after a cold winter. Tjøme is a summer haven for city dwellers, and it is where Oslo's mercantile elite had built cottages along the shore one hundred years ago. The cottages were small and primitive but were very close to the water. New generations subsequently took over the cottages and wanted to make them more comfortable and much bigger.

The social democrats who ruled Norway for several decades decided that the shoreline should be for everybody, not only for privileged and rich people. The original summerhouses, even if they block the shoreline for passing individuals, are legal and remain. Owners may maintain and renew the buildings but may not carry out expansion works. There is an absolute ban on new summerhouses, cottages, boathouses, and docks in the one-hundred-meter belt along the water.

However, as presented by Dagens Næringsliv—the Norwegian equivalent of the Wall Street Journal or the Financial Times—in the autumn of 2017, "cottages" along the coast had increased in size from 50 square meters and less to 200 square meters and more (Berglihn, 2019; DN, 2019). A cottage is normally small, while a summerhouse is normally larger. Two hundred square meters and more implies that there are often several bathrooms and many bedrooms. The largest ones can deserve the label of summer palaces and mansions. The rich and mighty had builders do blast in the mountains and remove rocks. They put up fences and created private parks so that nobody would dear to trespass, although it is legal to pass people's houses along the shoreline.

At the center of attention of the illegal construction and expansion of summer palaces, were an architect and a community planner in the municipality. The municipal planner had approved absolutely all of the applications from the architect, even those in the one-hundred-meter belt, and for a while, the community planner found convenient employment in the architectural firm before returning to the municipality.

Suspicions of corruption and financial adultery emerged. However, in this chapter, we are focusing on the white-collar offenders who hired architects and attorneys to make their cottages develop into summer-houses and summer palaces along the shore.

Two whistleblowers who live on the island in the summertime contacted a journalist at Dagens Næringsliv in the spring of 2017. They did not only present accusations. They had retrieved documents from the municipality for sixty key cottage cases. They presented the cases in a systematic way on a digital platform where journalists could conduct searches. Journalists quickly found names of owners who typically had received public attention before, either because they had inherited large sums of money or because they were self-made rich people. The media published pictures of the newly built summerhouses and summer palaces with pictures of the owners as well.

At first, local police were reluctant to look at the case. They probably knew that house owners with almost unlimited resources would do anything to avoid the demolition of their palaces. The house owners might easily be willing to pay fines, but the level of public outrage was high, since rich people had done things that normal people were unable to do. Therefore, the public request was to have all new palaces demolished.

It is interesting to shed some light on rich people's motives, opportunities, and behaviors as defined in the theory of convenience (Braaten & Vaughn, 2019; Chan & Gibbs, 2020; Dearden & Gottschalk, 2020; Gottschalk, 2017, 2019). Their motive in building illegal summer-houses is to climb Maslow's (1943) hierarchy of needs. The psychologist Abraham Maslow developed a hierarchy of human needs. This starts at the bottom with physiological needs, need for security, social needs, and need for respect and self-realization. When basic needs such as food and shelter are satisfied, then the person moves up the pyramid to satisfy needs for safety and control over their own life situation. Higher up in the pyramid, the person strives for status, recognition, and self-respect (Blickle et al., 2006). While street crime is often concerned with the lower levels, white-collar crime is often concerned with the upper levels in terms of status and success. Poverty and powerlessness tend to be the cause of street crime, while excessive power and greed are often the cause of white-collar crime.

Opportunity for white-collar offenders is concerned with access to resources to both commit and conceal crime (Benson & Simpson, 2018; Cohen & Felson, 1979). An opportunity is attractive as a means of responding to desires, wishes, and ambitions. The resource dimension provides the white-collar criminal with an opportunity to commit financial crime and conceal it in legal activities as a seemingly legitimate enterprise. Aguilera and Vadera (2008: 434) describe a criminal opportunity as "the presence of a favorable combination of circumstances that renders a possible course of action relevant". Opportunity arises when individuals or groups can engage in illegal and unethical behavior and expect, with a reasonable degree of confidence, to avoid detection and punishment. It is convenient for the offender to conceal crime and give it an appearance of outward acceptability.

The third and final dimension of convenience theory—in addition to motive and opportunity—is willingness for deviant behavior. Most theories concerned with fraud and corruption belong to the behavioral dimension. Researchers have presented numerous suggestions to explain why famous people are willing to commit white-collar crime despite holding privileged positions. Deviant behavior is an individual action that transgresses social norms, including formally enacted rules and informal nonconformity (Aguilera et al., 2018). The most frequently cited theory is neutralization, where offenders apply a number of techniques to rationalize and justify their deviant actions (Sykes & Matza, 1957). The idea of neutralization techniques resulted from work on Sutherland's (1939, 1983) theory of differential association. Criminals apply certain techniques in order to make them feel as though they have not done anything wrong. These guilt removal procedures are neutralization techniques, whereby the feeling of guilt decreases and possibly disappears. By neutralizing guilt feelings, offenders do not feel accountable, ashamed, or responsible (Chen & Moosmayer, 2020).

Evidence from attorneys' letters to the municipality as well as newspaper coverage of several cases suggests a number of neutralization techniques applied by homeowners. First, complexity prevented compliance (Lehman et al., 2019). Next, the rich people did not know what architects and construction firms did for them, so they deny responsibility (Sykes & Matza, 1957). Furthermore, some suggest normality of action, and some claim entitlement to action as some rich people think it is just a matter of price to obtain what you want.

Convicted White-Collar Offender

The convicted white-collar offender presented in the following is not a summerhouse owner on the island of Tjøme close to the city of Tønsberg. She has her summerhouse further south on the Norwegian coast close to the city of Grimstad. She is the first convicted environmental criminal in Norway for violations of the 100-meters belt regulations. Agder court of appeals wrote in its verdict (Agder, 2019: 20):

> There is a strong development pressure in the beach zone and individuals spend very high amounts of money to gain access to this limited good. Effective enforcement of established rules and guidelines is therefore necessary, and material and serious violations must result in tangible reactions. Often, purely economic reactions in terms of fines will have less deterrent effect.

This is interesting, as fines so far was the rule in Norway. However, rich people have no problems paying substantial fines, and then they tend to continue as before. Going to jail is another matter (Dhami, 2007; Logan, 2015; Logan et al., 2019; Stadler et al., 2013). The white-collar crime convict, the rich heir Hanne Madsen, received a verdict of 45 days in jail by the court.

Hanne Madsen had taken over a business enterprise from her father. She had also inherited the family summerhouse. She bought a property behind the summerhouse, which was within the one-hundred-meter belt as well. When acquiring the land, she told the municipality that she would do nothing to it, so that the public could continue walking across into a public beach nearby.

However, few years later, she had expanded the summerhouse and an annex. She had built an underground hallway from the summerhouse to the annex. On the acquired property, she built a tennis court.

The leading Norwegian business newspaper Dagens Næringsliv wrote in its editorial (DN, 2019: 2):

> This case is particularly serious, therefore a severe punishment. The violations of the Planning and Building Act included, among other things, blasting and construction of a 500 square meter concrete deck with artificial turf for a tennis court at the newly expanded summerhouse. A violation the municipality found that the owner had not even attempted or bothered to apply for a permit. The summerhouse owner has defended herself with a lack of knowledge of the illegal construction work. That may well be the case, but ignorance does not relieve guilt in the opinion of the law.

In court, Hanne Madsen presented herself almost as a victim of law violations, as she had trusted the architect, the attorney, the carpenter, the plumber, and all others involved in developing her vacation resort. Being a victim is a neutralization technique where the offender has a perception of victimization by an incident where others have ruined her life. The perception of being a victim increases as the media pressure and coverage becomes more personal.

Figure 4.1 emphasizes the likely convenience themes in the case of Hanne Madsen and rich and mighty individuals on the shoreline in general. In the motive dimension, there were no threats but rather possibilities at the individual level for the offender.

In the opportunity dimension, status and access to resources enabled the expansion of her vacation palace including a tennis court and an underground hallway. To the extent her architect and her attorney communicated with the local municipality, the communication served mainly the purpose of avoiding refusal or other problems for Madsen. Since a local community in Norway often suffers from institutional deterioration and lack of oversight, Madsen's helpers seem to have had an easy task. In the willingness dimension, Madsen felt no guilt as she blamed her helpers completely.

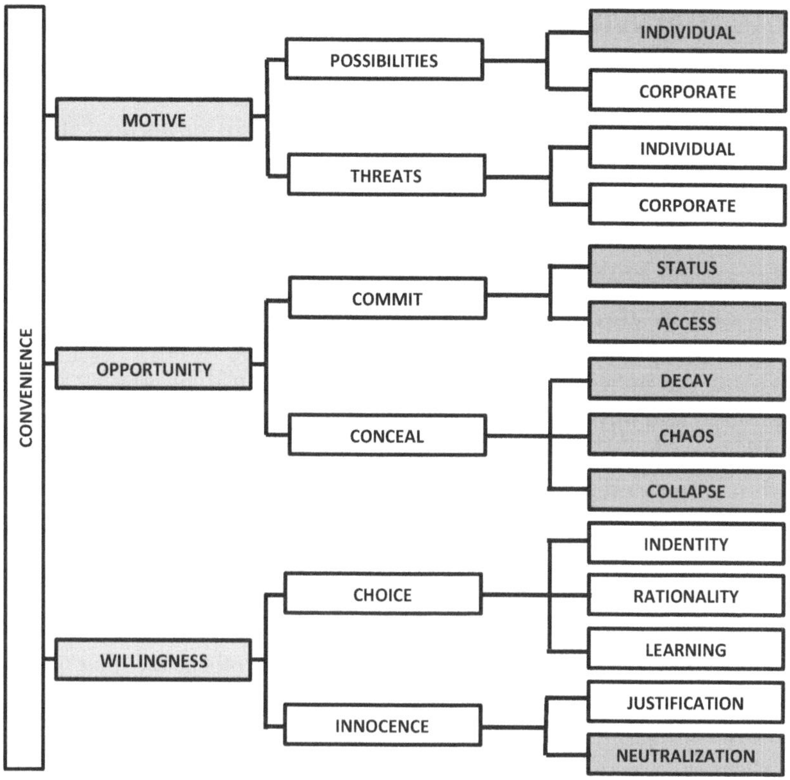

Fig. 4.1 Convenience themes in the case of a convicted summerhouse owner

It may seem potentially important that the offender sentenced in this case was a woman, which raises questions about why her and why at this particular time. After all, critics of Martha Stewart's case in the United States noted her seemingly overly harsh treatment in comparison with the many and much more serious harm caused by rich, white men that went and go unpunished.

Potential White-Collar Suspects

The Norwegian National Authority for Investigation and Prosecution of Economic and Environmental Crime (Økokrim) prosecuted the case against Hanne Madsen in court. This agency has the role of prosecuting larger and more serious cases that are matters of principle, while police districts prosecute common financial crime cases. The agency has an enforcement record of 90%, which implies that nine out of ten defendants end up in prison. This enforcement record is far above similar records in police districts. Økokrim is the main source of specialist skills for the police and prosecuting authorities in their combat against economic and environmental crime. Økokrim was established in 1989, and it is both a police specialist agency and a public prosecutor's office with national authority.

The district court emphasized the deterrence effect of its verdict (Agder, 2019: 20):

> A common measure of punishment for all offenses is strong general preventive considerations. It is important that others see that such violations cause strict reactions when they are revealed and prosecuted.

The sentence seems symbolic in relation to other cases typically handled by Økokrim. The national police agency Økokrim was thinking of cases against more summerhouse owners as well as architects and developers, and community planners as municipal government officials.

Figure 4.2 illustrates the situation on the island of Tjøme in terms of actors involved in potential wrongdoing and crime on the shoreline at the time of writing this chapter. Summerhouse owners were only one of the targets for police investigation and public prosecution. Other potential white-collar offenders were municipal politicians, municipal management, architects, and community planners, as well as firms that carried out construction work on the shoreline.

Once Økokrim had made their decisions regarding offenders and crimes, the suspects could expect charges. However, the suspects would then immediately hire the best (and most expensive) attorneys who would enter into symbolic defense and information control on behalf

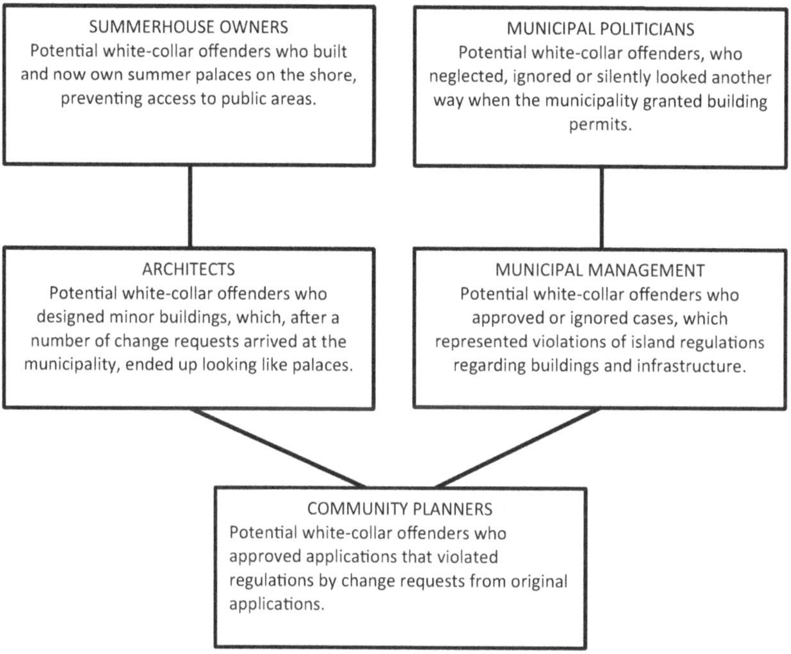

Fig. 4.2 Potential suspects in the Tjøme case for police investigators

of their clients. Therefore, police investigators from Økokrim and local police wanted to delay the indictments for as long as possible.

The most resourceful players in the Tjøme case—the summer palace owners—would very likely engage the nation's most skilled defense lawyers to assist them as victims of municipal wrongdoing and police misconduct. The country's top team of defense lawyers in white-collar cases is exceptionally skilled and thus terribly annoying for the police in their investigations. Their skills are thus possibly a reason why the police had not targeted more offenders in the initial phases of the investigation.

There were several possible hypotheses as to what kind of financial crime could be involved in the case. The possibilities are many in the Norwegian Penal Code: public fraud, tax evasion, incorrect company information, accounting fraud, public corruption, financial adultery, or money laundering, in addition to violations of building laws and environmental legislation.

In conclusion, this chapter has presented itself as a case study exploring illegal construction and expansion of a shoreline home in Norway, which led to jail time for the offender. The chapter argued that only the rich and mighty people commit these types of offenses since they have access to resources that allow them to circumvent the law. Relatedly, the jailing of the offender made this case unique and worthy of study. The case was also worthy of study because this area of research into environmental crime and justice tends to be marginalized in criminology and public policy. The chapter is the first criminological inquiry into the particular topic of the rich and mighty on the shoreline, which is indeed a challenge for the preservation of public access to attractive recreational resorts in many parts of the world.

This chapter has again applied the theory of convenience to white-collar offenses on the shoreline in Norway. The rich and mighty have occupied the most attractive recreational places along the coast with their vacation resorts in the form of summer palaces and private harbors. While Norwegian law makes it illegal to construct or expand buildings and other changes in the landscape in the one-hundred-meter belt along the coast, white-collar offenders hire architects, attorneys, and others to get what they want. For the first time in recent history, a white-collar offender received a jail sentence for such crime in Norway.

As an emerging explanation for the white-collar crime phenomenon, this study has illustrated convenience theory in the form of a relative concept where illegally building or expanding a summerhouse on the shoreline is relatively more convenient compared to not having such a mansion. The implication of this study is thus that convenience as a relative concept, where offenders have the choice between alternative actions, might lead to new insights into white-collar crime occurrences. We return to cases on the Tjøme Island in later chapters when describing the police investigation that followed.

Bibliography

Agder. (2019, September 16). *Court case 18-178766AST-ALAG*, Agder lagmannsrett (Agder court of appeals), Agder, Norway, 22 pp.

Aguilera, R. V., & Vadera, A. K. (2008). The dark side of authority: Antecedents, mechanisms, and outcomes of organizational corruption. *Journal of Business Ethics, 77,* 431–449.

Aguilera, R. V., Judge, W. Q., & Terjesen, S. A. (2018). Corporate governance deviance. *Academy of Management Review, 43*(1), 87–109.

BDO. (2017). *Tjøme commune – Gransking* [Tjøme municipality – Investigation], auditing firm *BDO*, Oslo, Norway, 39 pp.

Bell, C. M., & Khoury, C. (2016). Organizational powerlessness, dehumanization, and gendered effects of procedural justice. *Journal of Managerial Psychology, 31,* 570–585.

Benson, M. L., & Simpson, S.S. (2018). *White-collar crime: An opportunity perspective* (3rd ed.). Routledge.

Berglihn, H. (2019, September 20). *Princess-arving dømt til fengsel* [Princess heir sentenced to prison]. Daily Norwegian business newspaper *Dagens Næringsliv*, Friday, p. 4.

Berglihn, H. (2020). *Hanne Madsen må i fengsel: -Viktig at det reageres strengt* [Hanne Madsen must go to jail: -It is important that the reaction is strict]. Daily Norwegian business newspaper *Dagens Næringsliv*. www.dn.no, published June 29.

Blickle, G., Schlegel, A., Fassbender, P., & Klein, U. (2006). Some personality correlates of business white-collar crime. *Applied Psychology: An International Review, 55*(2), 220–233.

Braaten, C. N., & Vaughn, M. S. (2019). Convenience theory of cryptocurrency crime: A content analysis of U.S. federal court decisions. *Deviant Behavior*, published online https://doi.org/10.1080/01639625.2019.1706706.

Chan, F., & Gibbs, C. (2020). Integrated theories of white-collar and corporate crime. In: M. L. Rorie (Ed.), *The handbook of white-collar crime* (chapter 13, pp. 191–208). Wiley.

Chen, Y., & Moosmayer, D. C. (2020). When guilt is not enough: Interdependent self-construal as moderator of the relationship between guilt and ethical consumption in a confucian context. *Journal of Business Ethics, 161,* 551–572.

Cohen, L. E., & Felson, M. (1979). Social change and crime rate trends: A routine activity approach. *American Sociological Review, 44,* 588–608.

Cressey, D. (1972). *Criminal organization: Its elementary forms.* Harper and Row.

Dearden, T. E., & Gottschalk, P. (2020). Gender and white-collar crime: Convenience in target selection. *Deviant Behavior, Published Online.* https://doi.org/10.1080/01639625.2020.1756428.

Dhami, M. K. (2007). White-collar prisoners' perceptions of audience reaction. *Deviant Behavior, 28,* 57–77.

DN. (2019, September 21). *Ny dom kan skremme pengesterke hytteeiere fra ulovlig bygging – Ny dom gir god hjelp til kommuner som vil stoppe ulovlige byggearbeider i strandsonen* [New court sentence might scare money-rich cottage owners from illegal construction building – New court sentence provides good help to municipalities who want to stop illegal construction work in the beach zone]. daily Norwegian business newspaper *Dagens Næringsliv,* Saturday, p. 2.

Friedrichs, D. O. (2010). Integrated theories of white-collar crime. In F. T. Cullen & P. Wilcox (Eds.), *Encyclopedia of criminological theory* (Vol. 1, pp. 479–486). Sage.

Gibbs, C., & Boratto, R. (2017). Environmental crime. In H. N. Pontell (Ed.), *Oxford encyclopedia of criminology and criminal justice.* Published online oxfordre.com/criminology.

Gottschalk, P. (2017). Convenience in white-collar crime: Introducing a core concept. *Deviant Behavior, 38*(11), 605–619.

Gottschalk, P. (2019). *Convenience triangle in white-collar crime – Case studies of fraud examinations.* Edward Elgar Publishing.

Holm, R. C. (2020). To personer tiltalt for grov korrupsjon i hyttekommune [Two persons charged with serious corruption in cabin municipality]. Norwegian public radio *NRK Vestfold.* www.nrk.no, published April 30.

Holmøy, E. (2018). Byggesakslederen lot seg ikke stoppe av lover og regler – fjernet svaberg, lagde superanneks og grillhus med kjøkken og bad [The building case leader did not let himself be stopped by laws and rules – removed rocks, built annex and grill house with kitchen and bathroom]. local Norwegian daily newspaper *Tønsbergs Blad.* www.tb.no, published December 20.

Karim, K. E., & Siegel, P. H. (1998). A signal detection theory approach to analyzing the efficiency and effectiveness of auditing to detect management fraud. *Managerial Auditing Journal, 13*(6), 367–375.

Klevstrand, A. (2018, Jule 20). Nedgraderer Tjøme-arkitekt [Downgrade Tjøme architect]. Norwegian daily business newspaper *Dagens Næringsliv*, Wednesday, p. 17.

Klevstrand, A., and Solem, L. K. (2020, May 2). Økokrim undersøker hytteeiere, tiltaler arkitekt [Økokrim investigates cabin owners, charge architect]. Norwegian daily business newspaper *Dagens Næringsliv*, Saturday, pp. 12–13.

Krippendorff, K. (1980). *Content analysis: An introduction to its methodology*. Sage.

Krohn, M. D., & Eassey, J. M. (2014). Integrated theories of crime. In J. M. Miller (Ed.), *The encyclopedia of theoretical criminology* (pp. 458–463). Wiley.

Lehman, D. W., Cooil, B., & Ramanujam, R. (2019). The effects of rule complexity on organizational noncompliance and remediation: Evidence from restaurant health inspections. *Journal of Management*. published online, pp. 1–33. https://doi.org/10.1177/0149206319842262.

Lien, A. O., & Aunemo, E. W. (2019). Hun får betalt av fleire som er dypt involvert i Tjøme-skandalen. Nå har hun samlet byggebransjen mot kommunen [She gets paid by many who are deeply involved in the Tjøme scandal. Now she has gathered the construction business against the municipality]. daily Norwegian local newspaper *Tønsbergs Blad*. www.tb.no, published May 24.

Liska, A. E., Krohn, M. D., and Messner, S. F. (1989). Strategies and requisites for theoretical integration in the study of crime and deviance. In S. F. Messner, M. D. Krohn, & A. E. Liska (Eds.), *Theoretical integration in the study of deviance and crime* (pp. 1–20). State University of New York Press.

Logan, M. W. (2015). *Coping with imprisonment: Testing the special sensitivity hypothesis for white-collar offenders*. A dissertation to the Graduate School of the University of Cincinnati in partial fulfillment of the requirements for the degree of Doctor of Philosophy in the Department of Criminal Justice, Cincinnati, Ohio.

Logan, M. W., Morgan, M. A., Benson, M. L., & Cullen, F. T. (2019). Coping with imprisonment: Testing the special sensitivity hypothesis for white-collar offenders. *Justice Quarterly, 36*(2), 225–254.

Maslow, A. H. (1943). A theory of human motivation. *Psychological Review, 50*(4), 370–396.

McClelland, P. L., Liang, X., & Barker, V. L. (2010). CEO commitment to the status quo: Replication and extension using content analysis. *Journal of Management, 36*(5), 1251–1277.

Patrucco, A. S., Luzzini, D., & Ronchi, S. (2017). Research perspectives on public procurement: Content analysis of 14 years of publications in the Journal of Public Procurement. *Journal of Public Procurement, 16*(2), 229–269.

Pratt, M. G., Kaplan, S., & Whittington, R. (2019). Editorial essay: The tumult over transparency: Decoupling transparency from replication in establishing trustworthy qualitative research. *Administrative Science Quarterly*, 1–19. https://doi.org/10.1177/0001839219887663.

Saunders, M., Lewis, P., & Thornhill, A. (2007). *Research methods for business students* (5th ed.). Pearson Education.

Stadler, W. A., Benson, M. L., & Cullen, E. T. (2013). Revisiting the special sensitivity hypothesis: The prison experience of white-collar inmates. *Justice Quarterly, 30*(6), 1090–1114.

Sutherland, E. H. (1939). White-collar criminality. *American Sociological Review, 5*(1), 1–12.

Sutherland, E. H. (1983). *White collar crime – The uncut version.* Yale University Press.

Sykes, G. M., & Matza, D. (1957). Techniques of neutralization: A theory of delinquency. *American Sociological Review, 22*(6), 664–670.

Tilsynsrapport. (2020). *Gryteskjær fritidsbebyggelse tilsynsrapport* (Leisure Building Supervision Report). Færder kommune [Municipality of Færder]. reference 19/49893, dated January 17, 4 pp.

Wang, M. (2019). Mer trøbbel for hyttemillionær. Kommunen mener alle gamle tillatelser er ugyldige [More trouble for cabin millionaire: The municipality says that all old permits are invalid]. Norwegian local newspaper *Tønsbergs Blad*. www.tb.no, published December 5.

5
White-Collar Attorneys

The research topic of attorneys at work for white-collar offenders is traditionally associated with lawyers who help defendants while they are in the criminal justice system. During police investigation, court prosecution, and incarceration, an attorney can help the client in a subjective defense to limit the sanctions imposed on the client from the criminal justice system. The focus of research is then whether white-collar offenders are less likely to be incarcerated and receive shorter sentences compared to other offenders (Galvin, 2020), whether the special sensitivity hypothesis or the special resilience hypothesis applies to white-collar offenders (Logan et al., 2019), and other research topics related to law enforcement toward this group of offenders. In any event, standards of professional conduct indicate that lawyers are expected to zealously assert the client's position (Uribe et al., 2020).

P. Gottschalk, *Detecting and Investigating Environmental Crime*,
https://doi.org/10.1007/978-3-030-74184-6_5

Criminal Justice System Avoidance

This book applies a different perspective on attorneys at work for white-collar offenders. Inspired by Mann (1985), researchers have focused on attorneys at work for white-collar offenders who attempt to avoid the criminal justice system for their clients. Their approach is to describe deviance but deny crime. This approach is visible in a number of fraud examination reports, where attorneys have conducted private policing in client organizations. Client organizations hire attorneys from law firms to reconstruct past events and sequences of events that have caused scandals and crisis (Brooks & Button, 2011; Brooks et al., 2009; Button, 2019; Button & Gee, 2013; Button, Frimpong et al., 2007; Button, Johnston et al., 2007, Button et al., 2018; Gottschalk & Benson, 2020; King, 2020a, 2020b; Schneider, 2006; Williams, 2005a, 2005b, 2014). While attorneys conduct their investigations, the criminal justice system typically is reluctant to start public police investigations. When attorneys in their reports of investigations conclude that there was deviance but no crime, the criminal justice system typically remains reluctant to apply investigative resources to such cases (Gottschalk, 2019). Galvin (2020) suggests that the no-crime perspective results from "avoidance of the criminal justice system by institutionally framing 'law breaking' as '(regulatory) rule violations'." This phenomenon has received the label 'status advantage.'

It is important to emphasize that crime in criminology is a phenomenon with two characteristics. First, the action or non-action is bad, wrong, unethical, and immoral. Second, the action or non-action deserves a negative reaction. Therefore, crime is behavior that people condemn and that requires sanctions against the responsible person or entity.

Crime in legal terms is a phenomenon that penal law covers, where the law defines punishment for the behavior. If there is no law covering the deviant behavior, then it is no crime in a legal sense. Laws vary between jurisdiction, and laws change over time within each jurisdiction. It is the legal definition, rather than the criminological definition, that causes a suspect potentially to face the criminal justice system.

In criminology, deviance is negative behavior that transgresses accepted norms and expectations of other people, organizations, and society. Depending on the penal law coverage, the deviance may or may not be punishable.

Attorneys are knowledge workers who are supposed to be competent in general legal principles and procedures and in the substantive and procedural aspects of the law. They are supposed to have the ability to analyze and provide solutions to legal problems. Attorneys are professionals who have gained knowledge through formal education (explicit) and through learning on the job (implicit). An attorney is a knowledge worker specializing in the development and application of legal knowledge to solve client problems. Attorneys represent their clients in legal matters by presenting evidence and legal arguments as well as providing counsel to clients concerning their rights and obligations.

In white-collar research, there are different kinds of conception and operationalization of the construct of white-collar crime. Even though Sutherland (1939) coined the term more than eight decades ago, white-collar research remains preoccupied with definition. While many still engage in agitation and dispute about the white-collar crime concept, as exemplified by Galvin (2020), this research simply applies the basic characteristics of the offender. These characteristics include hierarchical status, personal respectability, individual trust, admiration by others, and legitimate access to resources, power base, and privileges. This implies that crime is committed in the course of occupational activities where the offender has legitimate access to premises and systems.

Rich and Mighty on the Shoreline

Status, respectability, trust, social esteem, access to resources, power, and privilege can expand beyond occupational activities into private activities for white-collar offenders. Access to attorneys at work can enable privileged individuals to explore and exploit possibilities that other people cannot do. While case studies so far in this book have focused on misconduct and wrongdoing in occupational settings, the following case study focuses on misconduct and wrongdoing in private settings by members of the elite in society.

This case study applies the theory of convenience to white-collar offenses on the shoreline in Norway. The rich and mighty are occupying the most attractive recreational places along the coast with their vacation resorts in the form of summer palaces and private harbors. This case study addresses the abuse of power associated with land development in an extremely attractive resort area in southern Norway. While Norwegian law makes it illegal to construct or expand buildings and other changes in the landscape in the one-hundred-meter belt along the coast, white-collar offenders hire architects, attorneys, and others to get what they want and defend what they have done.

Ever since Kenneth Mann published his portrait of attorneys at work (Mann, 1985), prosecutors and criminologists have noticed the special characteristics of defending white-collar wrongdoings. Attorneys do not limit their white-collar work to substance defense. They are also active in symbol defense and information control (Gottschalk, 2014). This chapter presents a case study of attorneys' white-collar work in support of offenders who have caused environmental harm on the shoreline in Norway.

The shoreline is a limited public good that many nations protect and preserve for the recreation and benefit of all its inhabitants and visitors. Privatization of the shoreline, especially in the form of vacation resorts, is thus illegal. Illegal construction in the coastal zone leads to irreversible interference with nature and restricts public access to coastal traffic. The beach zone is an area that is a vulnerable recreation place for the public in general, and an area under great development pressure with the possibility of great financial gain. Environmental crime combines with financial crime. Illegal private activity represents irreversible interventions harming the diversity of nature. Nevertheless, some rich and mighty people are able to violate the law. They build and expand their vacation homes on the shoreline by blasting mountain areas and preventing beach access. With the help of architects, builders, attorneys, and relationships to community planners, they use their resources to get and keep what they want.

While Norway has a very long coastline, the attractive real estates for vacation houses and summer guests are located south of the capital of Oslo. In particular, the island of Tjøme is popular (BDO, 2017; DN, 2019; Wang, 2018; Ytreberg, 2018). The theme of environmentally harmful construction and expansion of summerhouses serves as an example in relation to convenience theory. The motive is often to climb Maslow's (1943) hierarchy of needs, the opportunity is typically access to resources (Huisman & Erp, 2013), and the willingness for deviant behavior is dependent on neutralization techniques (Sykes & Matza, 1957), lack of self-control (Craig & Piquero, 2016), and other factors.

This chapter thus applies the theory of convenience to the case study. Convenience theory suggests that there is a motive, an opportunity, and a willingness among white-collar offenders to violate the law (Braaten & Vaughn, 2019; Chan & Gibbs, 2020; Dearden & Gottschalk, 2020; Gottschalk, 2017, 2019). The theory builds on the fraud triangle suggested by Cressey (1972). A white-collar offender is a member of the upper echelon in society who abuses the privileged position for illegitimate benefits (Sutherland, 1939, 1983).

In white-collar research, there are different kinds of conception and operationalization of the construct of white-collar crime. Even though Sutherland (1939) coined the term more than eight decades ago, white-collar research remains preoccupied with definition. While many still engage in agitation and dispute about the white-collar crime concept, as exemplified by Galvin (2020), this research simply applies the basic characteristics of the offender that include status, trust, and access, as discussed previously in this book.

Environmental harm and crime has received increased attention in recent years (Böhm, 2020; Huisman & Erp, 2013; Lynch, 2020). Traditionally, white-collar crime cases have focused on non-violent financial crime. Recently, with increased environmentalism, researchers have focused on white-collar crime that can impose physical harm on people (Benson & Simpson, 2018: 129):

> These offenses are potentially much more serious in that they can and often do impose physical costs on individuals. This is not to say that the perpetrators deliberately set out to harm other people. They do not. The

> physical harms that they cause are unintended in the sense that they are not what the offender is trying to achieve. The motivation for the offense is not to impose harm on others but rather to gain a financial advantage.

For example, Wingerde and Lord (2020: 478) argue that the waste industry is a criminogenic industry that is vulnerable to environmental crime:

> First, this concerns the waste product itself. Waste is a product that has a negative value attached to it (…). Second, the industry in itself also has some characteristics that are considered to be criminogenic.

However, few individuals face convictions for environmental crime. For example, after the British Petroleum (BP) Deepwater Horizon oil spill in the Gulf of Mexico in 2010, prosecutors brought criminal charges against four British Petroleum executives, but no one ended up in prison (Fowler, 2014; Freeh, 2013; Thompson, 2017). Greife and Maume (2020) found that:

> Despite recent attention to multi-billion-dollar settlements for environmental violations involving high-profile offenders such as BP and VW, criminal sanctioning of individuals and organizations for environmental offenses is uncommon.

Going to prison is a serious matter (Dhami, 2007; Logan et al., 2019; Stadler et al., 2013). One exception is the conviction of both the chief executive officer and the chairperson of the board at Vest Tank in Norway, as well as a chemical advisor to the company, who received prison sentences for a tank explosion caused by dangerous waste (Gormley, 2009; Gulating, 2013, 2015; Nordhordland, 2010; Pedersen, 2017).

The environmental harm in the current case study includes illegal changes in the landscape, pollution from civilization, conquest, and privatization of a public good, and prevention of recreational opportunities for people.

Convenient Loopholes for Mansions

Convenience theory is an emerging approach to integrate well-known perspectives on white-collar crime from criminology, sociology, and management. It is an integrated deductive theory (Chan & Gibbs, 2020; Friedrichs, 2010; Krohn & Eassey, 2014; Liska et al., 1989; Pratt et al., 2019). The choice of convenience theory in this book does not result from a comparison to other perspectives such as power or class. Rather, as an emerging new theoretical perspective on white-collar offenders, it is interesting to explore its empirical application to a case of environmental crime.

Convenience theory suggests that the extent of convenience determines the likelihood of white-collar wrongdoing, which is economic crime committed by privileged individuals with access to resources, who abuse their positions for illegitimate benefits (Piquero, 2018; Sutherland, 1939, 1983). Convenience is the state of being able to proceed with something with little effort or difficulty, avoiding pain and strain (Mai & Olsen, 2016). Convenience is savings in time and effort (Farquhar & Rowley, 2009), as well as avoidance of pain and obstacles (Higgins, 1997). Convenience is a relative concept concerned with efficiency in time and effort as well as reduction in pain and solution to problems (Engdahl, 2015). Convenience is an advantage in favor of a specific action to the detriment of alternative actions. White-collar offenders choose the most convenient path to reach their goals (Wikstrom et al., 2018).

A convenient individual is not necessarily neither bad nor lazy. On the contrary, the person can be seen as smart and rational (Sundström & Radon, 2015). Convenience orientation varies among individuals, as some are more concerned about time saving, effort reduction, and pain avoidance and thus more convenience oriented. For example, Gamache and McNamara (2019: 923) suggest that offenders in the upper echelon who mainly focus on their present time dispositions in their information selection are more inclined to justify their wrongdoing since they have "the proclivity to act impulsively, move quickly with current opportunities, and consider current issues when making decisions."

The theory of convenience defines three dimensions that influence the tendency of white-collar occurrences. First in the motive dimension, crime is likely to occur if it is possible to avoid threats conveniently, and if it is feasible to explore possibilities conveniently. Examples of threats include falling from position (Piquero, 2012) and bankruptcy (Blickle et al., 2006), while examples of possibilities include climbing in the hierarchy of needs (Agnew, 2014; Maslow, 1943) and satisfying greed (Goldstraw-White, 2012). Greed leads to a need for an increasingly larger home, several chalets and summerhouses, bigger boats, luxurious vacations, and ownership in various enterprises. Greed is a desire among all sorts of people. When there are convenient possibilities for financial gain to enjoy prosperity, then economic crime can be a convenient action. Both Bucy et al. (2008) and Hamilton and Micklethwait (2006) emphasize greed as the most common cause of criminal acts by white-collar offenders.

Acquiring a summer castle on the shoreline is a typical example of climbing in the hierarchy of needs, where only rich and mighty individuals are able to buy or build such vacation resorts (Ytreberg, 2018). Norwegians love the outdoors, and many families have a cabin in the mountains for skiing in the winter, as well as a summerhouse close to the ocean for swimming and boat trips in the summer. The most attractive locations are extremely expensive, and many areas along the shoreline are exclusively for the public in general.

Second, in the opportunity dimension of convenience theory, if it is convenient to commit as well as to conceal illegitimate activity, then wrongdoing is likely to occur. Examples of convenience in committing crime include the too-big-to-fail and too-big-to fall phenomenon (Pontell et al., 2014), and legitimate access to resources (Benson & Simpson, 2018; Cohen & Felson, 1979; Logan et al., 2019). Examples of convenience in concealing crime include institutional deterioration (Rodriguez et al., 2005), lack of controls (Bosse & Phillips, 2016), and rule complexity preventing compliance (Lehman et al., 2019). A rich and mighty person has legitimate access to architects and builders to plan and construct a summerhouse. The person also has access to attorneys and others who can negotiate and influence municipal officials to get approvals or to avoid refusals because of rule complexity in the legislation (Lehman et al., 2019).

Attorneys do not only provide substance defense for their clients, they often also contribute symbolic defense and information control to manipulate community planners in the municipalities (Gottschalk, 2014). The environmental harm and thus the wrongdoing can disappear from sight, for example by first removing rock and then building large basements that are not visible to trespassers.

To seek attractive and exclusive homeownership in the most convenient way is not necessarily to choose a new construction or extension in such a modality that would require rich and mighty individuals to hire highly specialized and deviant contractors and attorneys. Rather, the chosen approach is sometimes to find loopholes so that the emerging and expanding mansion at the shoreline seemingly has transgressed no laws or regulations. Laws and regulations tend to be general in their formulations, as the text is supposed to cover all conceivable forms of nonconformity presently and in the future.

Finally, in the willingness dimension, if deviant behavior is a convenient choice where the feeling of justification and innocence is present, then crime is likely to occur. Examples of convenience in choice of crime include deviant identity (Obodaru, 2017; Petrocelli et al., 2003), narcissistic identification with the location (Galvin et al., 2015), perception of benefits exceeding costs (Pratt & Cullen, 2005), and learning from others by differential association (Sutherland, 1983). Examples of convenience by innocence include justification of wrongdoing (Schnatterly et al., 2018) as well as application of neutralization techniques (Sykes & Matza, 1957). A rich and mighty person building an illegal vacation resort on the shoreline can apply, for example, the neutralization technique of denial of responsibility. The person disclaims responsibility for crime, as the person is not the correct receiver of blame for what happened, since an entrepreneur handled the project. The offender here claims that one or more of the conditions of responsible agency did not occur. The person committing a deviant act defines himself or herself as lacking responsibility for his or her actions. In this technique, the person rationalizes that the action in question is beyond his or her control.

The rich and mighty person claims that architects, builders, and attorneys handled the matter on his or her behalf where agency problems occurred. In the principal-agent perspective, the elite member can claim as the principal that the architect as an agent had a greater risk willingness and a deviant priority compared to the agent, where the agent had no understanding of what was going on because of knowledge asymmetry between principal and agent (Bosse & Phillips, 2016).

Repeated figures in this book have illustrated the structure of white-collar crime convenience. The extent of white-collar convenience manifests itself by motive, opportunity, and willingness. In the motive dimension, wealth might be a goal in itself or an enabler to exploit possibilities and to avoid threats. In the current case study, environmental crime enables climbing in the hierarchy of needs by building or expanding a vacation resort on the shoreline. Generally, it is convenient to exploit possibilities and to avoid threats by financial means (Agnew, 2014; Blickle et al., 2006; Goldstraw-White, 2012; Maslow, 1943; Piquero, 2012, 2018).

In the opportunity dimension, convenience can exist both to commit white-collar crime and to conceal white-collar crime. Offenders have high social status in privileged positions, and they have legitimate access to crime resources. In the current case study, wealthy offenders were able to hire architects, builders, and attorneys to get and keep what they. Generally, disorganized institutional deterioration causes decay, lack of oversight and guardianship causes chaos, and criminal market structures cause collapse (Benson & Simpson, 2018; Cohen & Felson, 1979; Gottschalk, 2017; Pontell et al., 2014; Qiu & Slezak, 2019).

The personal willingness for deviant behavior manifests itself by offender choice and perceived innocence. The choice of crime can be caused by deviant identity, rational consideration, or learning from others. The perceived innocence at crime manifests itself by justification and neutralization. Identity, rationality, learning, justification, and neutralization all contribute to making white-collar wrongdoing a convenient behavior for offenders (Craig & Piquero, 2017; Engdahl, 2015; Pratt & Cullen, 2005; Sutherland, 1983; Sykes & Matza, 1957).

Elite Mobilization of Top Attorneys

This research is concerned with application of convenience theory to the case of rich and mighty white-collar offenders who commit environmental crime by privatizing and harming the shoreline in terms of environmental wrongdoing (Olaussen, 2018; Ytreberg, 2018). Specifically, this research attempts to explore the opportunity dimension of convenience theory as illustrated repeatedly in this book, where attorneys can help commit and conceal environmental harm. White-collar offenders pay attorneys for their substance defense, symbolic defense and information control.

This archival research is based on content analysis of newspaper reports (e.g., Aunemo, 2018, 2020; Dugstad & Skaalmo, 2017; Holm, 2019, 2020; Holmøy, 2017, 2018, 2019, 2020; Wang, 2018, 2019; Ytreberg, 2018), municipal files (e.g., Tilsynsrapport, 2020), fraud examinations (e.g., BDO, 2017), letters to the municipality of Tjøme from attorneys, and email communication and dialogues with whistleblowers. Whistleblowing is the disclosure by organizational members of illegal, immoral, or illegitimate practices under the control of their employers, to persons or organizations that may be able to effect action to terminate such practices (Wells et al., 2020). Content analysis is any methodology or procedure that works to identify characteristics within texts attempting to make valid inferences (Krippendorff, 1980; Patrucco et al., 2017). Content analysis assumes that language reflects both how people understand their surroundings and their cognitive processes (Bell et al., 2018; Saunders et al., 2007). Therefore, content analysis makes it possible to identify and determine relevant text in a context (McClelland et al., 2010).

In addition to archival material, a whistleblower continued his search for cases and communicated his finding to this research effort. He visited potential crime scenes, took pictures, and continued his dialogue with both politicians and staff in the municipality.

Over a period of three years, a substantial stream of emails and personal communications enabled this research. Thirty-six white-collar cases caused offenders to hire a number of attorneys as listed in Table 5.1 over the three-year period 2017–2020 as the whistleblowers had attracted

Table 5.1 Identified attorneys in law firms and their white-collar clients

	Attorney	Law firm	White-collar client
1	Andresen, Ole	Riisa	Ole Andresen
2	Bech, Stig L.	BAHR	Rune Breili
3	Blaauw, Kristine R.	Thommessen	Arnfinn Hafsteen
4	Borge, Ole	Borge	Ole Borge
5	Bruusgaard, Erik	Tenden	Arnfinn Hafsteen
6	Bæra, Vibeke Hein	Hei Bæra	Rune Breili
7	Bæra, Vibeke Hein	Hein Bæra	Werner Osther
8	Eilertsen, Eilert	Byadvokaten	Nanna Fløgstad Sohlberg
9	Gulsvik, Mari Helen	Selmer	Gunnar Fredrik Selvaag
10	Hansson, Nils Petter	Ro Sommernes	Hans Ole Helling
11	Harborg, Henning	Thommessen	Carl Erik Krefting
12	Harborg, Henning	Thommessen	Turid Varner
13	Hauger, Bård Bergem	Advokatene på Jeløy	Hans Petter Abrahamsen
14	Jørgensen, Frode S.	Sandvika advokatkontor	Harald Dybsjord
15	Jørgensen, Frode S.	Sandvika advokatkontor	Sigmund Arnesen
16	Lie, Arne	Haraldsen Bydal Lie	Harald Svendsen
17	Nåmdal, Endre	Riisa	Rune Breili
18	Reusch, Christian H.P.	Simonsen Vogt Wiig	Kristina M. Wilhelmsen
19	Støren, Martin	Schjødt	Lars Hellik Strøm
20	Støren, Martin	Schjødt	Sigmund Arnesen
21	Støren, Martin	Schjødt	Tore Jacobsen
22	Sæveraas, Ingeborg	Lønnum	Øyvind Garner
23	Tellefsen, Anne	Haraldsen Bydal Lie	Fridtjov Botvid Falck
24	Tellefsen, Anne	Haraldsen Bydal Lie	Celestine Clauson
25	Tellefsen, Anne	Haraldsen Bydal Lie	Nanna Fløgstad Sohlberg
26	Tellefsen, Anne	Haraldsen Bydal Lie	Harald Svendsen

(continued)

Table 5.1 (continued)

	Attorney	Law firm	White-collar client
27	Tellefsen, Anne	Haraldsen Bydal Lie	77 construction firms
28	Tellefsen, Anne	Haraldsen Bydal Lie	Rune Breili
29	Tellefsen, Anne	Haraldsen Bydal Lie	Svein Homstvedt
30	Tellefsen, Anne	Haraldsen Bydal Lie	Scammer and scammed
31	Tellefsen, Anne	Haraldsen Bydal Lie	Malling og Midelfart
32	Tveten, Mathias	Ro Sommernes	Tom Ragnar Wilhelmsen
33	Waller, Caroline	Simonsen Vogt Wiig	Anders Wilhelmsen
34	Zimmermann, Liv	Hjort	Mette Landgraff
35	Zimmermann, Liv	Hjort	Evy Reimers
36	Zimmermann, Liv	Hjort	Marianne and Ulf Stuge

the attention of politicians and staff in the municipality toward rich and mighty persons on the shoreline (Brevik, 2020).

The elite mobilization of top attorneys was a result of media coverage and municipal wakening about the de facto closing of the shoreline by rich and mighty people who built and expanded their summer palaces along the shoreline. All attorneys started their substance defense, symbol defense, and information control as a consequence of being hired by the rich and wealthy when investigative journalists (Skavhellen & Lien, 2018), whistleblowers and some left-wing local politicians revealed summerhouse after summerhouse that did not have approval from the relevant authorities. The role of attorneys in these individual scandals is similar to their role in corporate scandals as discussed in this book, where they conclude with misconduct but no crime.

One important role of top attorneys is to cause delays when the rich and mighty are caught in the criminal justice system. An example is attorney Erling Høyte at law firm Falkanger, Wiik, Høyte & Fremstad on behalf of Bjørn Wahl who is a rich Norwegian citizen living in Switzerland. Already in 2014, the municipality of Nøtterøy detected an illegal wharf and harbor at Wahl's property. The municipality ordered Wahl to demolish the illegal harbor in 2015. Attorney Høyte sued the municipality in court in 2017. The district court acquitted the municipality in 2020 (Vestfold tingrett, 2020). It was expected that Bjørn Wahl would ask Erling Høyte to appeal the verdict. The likely delay from detection of wrongdoing to demolition of the harbor would probably become one decade.

The Island of Tjøme in Norway

Our case is concerned with illegally constructed summerhouses and expanded vacation resorts on the beaches and cliffs in the southern part of Norway. Both environmental protection laws and public access laws have made it illegal to build and expand vacation homes in the hundred meters wide belt along the coast.

Nevertheless, rich people have been able to ignore the building ban in the case of municipality of Tjøme (Aunemo, 2018, 2020; Heltne & Klevstrand, 2019; Holm, 2020; Holmøy, 2019; Larssen, 2017; Lien & Brevik, 2020; Wang, 2019). The environmental harm from shoreline violations is twofold. First, illegal privatization reduces and removes public access to attractive recreational areas. Public health suffers harm, and democracy where everyone is equal in front of the law, deteriorates. Second, construction work changes the landscape. Nature and animals suffer harm.

The municipality of Tjøme is located on an island south of the Norwegian capital, Oslo. The island is the most expensive summerhouse resort and may deserve a comparison to Cape Cod in the United States in terms of its attractiveness for people after a cold winter. Tjøme is a summer haven for city dwellers, and it is where Oslo's mercantile elite started building cottages along the shore one hundred years ago (Holth & Winge, 2018). The cottages were small and primitive but were very close to the water. New generations subsequently took over the cottages and wanted to make them more comfortable and much larger (Klevstrand & Randen, 2018).

The social democrats who ruled Norway for several decades decided that the shoreline should be for everybody, not only for privileged and rich people. The original summerhouses, even if they block the shoreline for passing individuals, are legal and remain. Owners may maintain and renew the buildings but may not carry out expansion work. There is an absolute ban on new summerhouses, cottages, boathouses, and docks in the one-hundred-meter belt along the waterfront.

However, as presented by Dagens Næringsliv—the Norwegian equivalent of the Wall Street Journal or the Financial Times—in the autumn of 2017, "cottages" along the coast had increased in size from 50 square meters and less to 200 square meters and more (Berglihn, 2019; DN, 2019). A cottage is normally small, while a summerhouse is normally larger. Two hundred square meters and more, such as three thousand square feet, imply that there are often several bathrooms and many bedrooms. The largest ones can deserve the label of summer palaces almost the size of the Vanderbilt mansion in Newport, Rhode Island in the United States.

The rich and mighty had builders do blast of the mountains and remove rocks (Blix, 2020). They put up fences into the water and created private parks so that nobody would dare to trespass, although it is legal to pass people's houses along the shoreline. In terms of crime signals (Karim & Siegel, 1998), the signal intensity from illegal construction sites was low, except when people observed construction work from boats on the sea. The local population had declining signal alertness, as they knew that no terminations or corrections would occur anyway.

At the center of attention of the illegal construction and expansion of recreational palaces were an architect and a community planner in the municipality of Tjøme (Blix, 2018; Dugstad & Skaalmo, 2017; Holm, 2019; Klevstrand, 2018; Klevstrand et al., 2020; Wang, 2018). The municipal planner had approved absolutely all of the applications from the architect, even those in the one-hundred-meter belt, and for a while, the community planner found convenient employment in the architectural firm with a nice salary before returning to the municipality. Suspicions of corruption and financial adultery emerged (Holmøy, 2017, 2018). However, in this chapter, we are focusing on the white-collar offenders who hired architects and attorneys listed in Table 5.1 to make their summer cottages develop into summerhouses, summer palaces, and summer mansions along the shoreline (Blix, 2019; Lien & Aunemo, 2019).

Two whistleblowers who live on the island contacted a journalist at Dagens Næringsliv in the spring of 2017. They did not only present accusations. They had retrieved documents from the municipality for sixty key summerhouse cases. They presented the cases in a systematic way on a digital platform where journalists could conduct searches.

Journalists quickly found names of owners who typically had received public attention before, either because they had inherited large sums of money or because they were self-made rich people. The media published pictures of the newly built summerhouses and summer palaces with pictures of the owners as well.

Local police were reluctant to look at the cases (Klevstrand & Solem, 2020). They probably knew that house owners with almost unlimited resources would do anything to avoid the demolition of their palaces. The house owners might easily be willing to pay fines, but the level of public outrage was high, since rich people had done things that normal people were unable to do. Therefore, the public request was to have all new palaces demolished.

It is interesting to shed some light on rich people's motives, opportunities, and behaviors as defined in the theory of convenience (Braaten & Vaughn, 2019; Chan & Gibbs, 2020; Dearden & Gottschalk, 2020; Gottschalk, 2017, 2019). Their motive in building illegal summerhouses is to climb Maslow's (1943) hierarchy of needs. The psychologist Abraham Maslow developed a hierarchy of human needs. This starts at the bottom with physiological needs, need for security, social needs, and need for respect and self-realization. When basic needs such as food and shelter are satisfied, then the person moves up the pyramid to satisfy needs for safety and control over their own life situation. Higher up in the pyramid, the person strives for status, recognition, and self-respect (Blickle et al., 2006). While street crime is often concerned with the lower levels, white-collar crime is often concerned with the upper levels in terms of status and success. Poverty and powerlessness tends to be the cause of street crime, while excessive power and greed is often the cause of white-collar crime.

Opportunity for white-collar offenders is concerned with access to resources to both commit and conceal crime (Benson & Simpson, 2018; Cohen & Felson, 1979). An opportunity is attractive as a means of responding to desires, wishes, and ambitions. The resource dimension provides the white-collar criminal with an opportunity to commit financial crime and conceal it in legal activities as a seemingly legitimate enterprise. Building and expanding mansions on the coast increase the value of properties far beyond the material costs. The offense is thus just as much economic crime as environmental crime (Holmøy, 2020).

Aguilera and Vadera (2008: 434) describe a criminal opportunity as "the presence of a favorable combination of circumstances that renders a possible course of action relevant." Opportunity arises when individuals or groups can engage in illegal and unethical behavior and expect, with a reasonable degree of confidence, to avoid detection and punishment. It is convenient for the offender to conceal crime and give it an appearance of outward acceptability.

The third and final dimension of convenience theory—in addition to motive and opportunity—is willingness for deviant behavior. Most theories concerned with fraud and corruption belong to the behavioral dimension. Researchers have presented numerous suggestions to explain why famous people are willing to commit white-collar crime despite holding privileged positions. Deviant behavior is an individual action that transgresses social norms, including formally enacted rules and informal nonconformity (Aguilera et al., 2018). The most frequently cited theory is neutralization, where offenders apply a number of techniques to rationalize and justify their deviant actions (Sykes & Matza, 1957). The idea of neutralization techniques resulted from work on Sutherland's (1939, 1983) theory of differential association. Criminals apply certain techniques in order to make them feel as though they have not done anything wrong. These guilt removal procedures are neutralization techniques, whereby the feeling of guilt decreases and possibly disappears. One out of several neutralization techniques available to mansion owners was that everyone else did what they do, nicely coordinated by a key architect who had convenient access to corrupt officials in the municipality.

Evidence from attorneys' letters to the municipality as well as newspaper coverage of several cases suggests a number of neutralization techniques applied by homeowners. First, complexity prevented compliance (Lehman et al., 2019). Next, the rich people did not know what architects and construction firms did for them, so they deny responsibility (Sykes & Matza, 1957). Furthermore, some suggest normality of action, and some claim entitlement to action as some rich people think it is just a matter of money to get what you want.

However, in the social democratic tradition in Norway, which is typically to the left on the political spectrum compared to Democrats in the United States, the state and the people are above the corporations and the individuals. The role of the state is to regulate businesses and behaviors. This is in contrast to some other countries, where the state primarily serves the corporate world and the freedom of individuals to do whatever they like, even when they harm the public good.

The theme of illegal construction and expansion of summerhouses thus serves as an example in relation to convenience theory, where the motive is to climb Maslow's hierarchy of needs, the opportunity is access to resources, and the willingness for deviant behavior is dependent on neutralization techniques, lack of self-control and other enablers of wrongdoing willingness.

Bibliography

Agnew, R. (2014). Social concern and crime: Moving beyond the assumption of simple self-interest. *Criminology, 52*(1), 1–32.

Aguilera, R. V., Judge, W. Q., & Terjesen, S. A. (2018). Corporate governance deviance. *Academy of Management Review, 43*(1), 87–109.

Aguilera, R. V., & Vadera, A. K. (2008). The dark side of authority: Antecedents, mechanisms, and outcomes of organizational corruption. *Journal of Business Ethics, 77,* 431–449.

Aunemo, E. W. (2018, September 12). Her får Tønsbergs Blad-journalistene pris for Tjøme-sakene [Here the Tønsberg Blad journalists receive a prize for the Tjøme cases], daily local Norwegian newspaper *Tønsbergs Blad*. www.tb.no, published Wednesday.

Aunemo, E. W. (2020). Advokaten til millionærarving Celestine (22) krever at skattebetalerne skal betale honoraret hennes [The millionaire heiress attorney demands that the taxpayers pay her fees], daily local Norwegian newspaper *Tønsbergs Blad*. www.tb.no, published January 19.

BDO. (2017). Tjøme commune – Gransking [Tjøme municipality – Investigation], auditing firm *BDO*, Oslo, Norway, 39 pp.

Bell, E., Bryman, A., & Harley, B. (2018). *Business research methods* (2nd ed.). Oxford University Press.

Benson, M. L., & Simpson, S. S. (2018). *White-collar crime: An opportunity perspective* (3rd ed.). Routledge.

Berglihn, H. (2019, September 20). Princess-arving dømt til fengsel [Princess heir sentenced to prison], daily Norwegian business newspaper *Dagens Næringsliv*, Friday, p. 4.

Berglihn, H. (2020). Hanne Madsen må i fengsel: -Viktig at det reageres strengt [Hanne Madsen must go to jail: -It is important that the reaction is strict], daily Norwegian business newspaper *Dagens Næringsliv*. www.dn.no, published June 29.

Berglihn, H., Gjernes, K., & Nordahl, B. O. (2007). *Metoderapport: Skjult i skatteparadis* [Method report: Hidden in tax haven], daily Norwegian business newspaper Dagens Næringsliv. www.dn.no, published January 18.

Blickle, G., Schlegel, A., Fassbender, P., & Klein, U. (2006). Some personality correlates of business white-collar crime. *Applied Psychology: an International Review, 55*(2), 220–233.

Blix, N. T. (2018). Økokrim får fortsatt mange Tjøme-saker og regner med at etterforskningen strekker seg langt ut i 2019 [Økokrim still receives many Tjøme cases and expect the investigation to last far into 2019], daily local Norwegian newspaper *Tønsbergs Blad*. www.tb.no, published November 9.

Blix, N. T. (2019). Rådmannen vil avslå eks-ordførerens søknad om bod i sjøkanten ved Mågerø: -Det innebærer en privatisering av arealene [The councilor will refuse the ex-mayor's application for a shed at the seafront on Måger island: -It implies the privatization of land], local newspaper *Øyene*. www.oyene.no, published October 27.

Blix, N. T. (2020). Saken er fortsatt ikke over: Tvangsmulkten på 20.000 kroner har vokst til 200.000 for ulovlig sand [The case is still not over: The coercive punishment of 20,000 kroner has grown to 200,000 for illegal sand], local newspaper *Øyene*. www.oyene.no, published April 7.

Bosse, D. A., & Phillips, R. A. (2016). Agency theory and bounded self-interest. *Academy of Management Review, 41*(2), 276–297.

Braaten, C. N., & Vaughn, M. S. (2019). Convenience theory of cryptocurrency crime: A content analysis of U.S. federal court decisions. *Deviant Behavior*, published online https://doi.org/10.1080/01639625.2019.1706706.

Brevik, H. C. (2020). Tom er ikke overrasket [Tom is not surprised], daily Norwegian local newspaper *Tønsbergs Blad*. www.tb.no, published April 30.

Brooks, G., & Button, M. (2011). The police and fraud investigation and the case for a nationalized solution in the United Kingdom. *The Police Journal, 84,* 305–319.

Brooks, G., Button, M., & Frimpong, K. (2009). Policing fraud in the private sector: A survey of the FTSE 100 companies in the UK. *International Journal of Police Science and Management, 11*(4), 493–504.

Bucy, P. H., Formby, E. P., Raspanti, M. S., & Rooney, K. E. (2008). Why do they do it? The motives, mores, and character of white collar criminals, *St. John's Law Review, 82,* 401–571.

Button, M. (2019). The "new" private security industry, the private policing of cyberspace and the regulatory questions. *Journal of Contemporary Criminal Justice*, published online https://doi.org/10.1177/1043986219890194.

Button, M., & Gee, J. (2013). *Countering fraud for competitive advantage – The professional approach to reducing the last great hidden cost*. Wiley.

Button, M., Frimpong, K., Smith, G., & Johnston, L. (2007). Professionalizing counter fraud specialists in the UK: Assessing progress and recommendations for reform. *Crime Prevention and Community Safety, 9,* 92–101.

Button, M., Johnston, L., Frimpong, K., & Smith, G. (2007). New directions in policing fraud: The emergence of the counter fraud specialists in the United Kingdom. *International Journal of the Sociology of Law, 35,* 192–208.

Button, M., Shepherd, D., & Blackbourn, D. (2018). 'The iceberg beneath the sea', fraudsters and their punishment through non-criminal justice in the 'fraud justice network' in England and Wales. *International Journal of Law, Crime and Justice, 53,* 56–66.

Böhm, M. L. (2020). Criminal business relationships between commodity regions and industrialized countries: The hard road from raw material to new technology. *Journal of White Collar and Corporate Crime, 1*(1), 34–49.

Chan, F., & Gibbs, C. (2020). Integrated theories of white-collar and corporate crime. In M. L. Rorie (Ed.), *The handbook of white-collar crime* (chapter 13, pp. 191–208). Wiley.

Cohen, L. E., & Felson, M. (1979). Social change and crime rate trends: A routine activity approach. *American Sociological Review, 44,* 588–608.

Craig, J. M., & Piquero, N. L. (2016). The effects of low self-control and desire-for-control on white-collar offending: A replication. *Deviant Behavior, 37*(11), 1308–1324.

Craig, J. M., & Piquero, N. L. (2017). Sensational offending: An application of sensation seeking to white-collar and conventional crimes. *Crime & Delinquency, 63*(11), 1363–1382.

Cressey, D. (1972). *Criminal organization: Its elementary forms*. Harper and Row.

Dearden, T. E., & Gottschalk, P. (2020). Gender and white-collar crime: Convenience in target selection. *Deviant Behavior*, published online https://doi.org/10.1080/01639625.2020.1756428.

Dhami, M. K. (2007). White-collar prisoners' perceptions of audience reaction. *Deviant Behavior, 28*, 57–77.

DN. (2019, September 21). Ny dom kan skremme pengesterke hytteeiere fra ulovlig bygging – Ny dom gir god hjelp til kommuner som vil stoppe ulovlige byggearbeider i strandsonen [New court sentence might scare money-rich cottage owners from illegal construction building – New court sentence provides good help to municipalities who want to stop illegal construction work in the beach zone], daily Norwegian business newspaper *Dagens Næringsliv*, Saturday, p. 2.

Dugstad, L., & Skaalmo, G. (2017). Slik fikk han de rikes hytter godkjent på Tjøme [This is how he got the cabins of the rich approved on Tjøme], daily Norwegian business newspaper *Dagens Næringsliv*. www.dn.no, published September 29.

Engdahl, O. (2015). White-collar crime and first-time adult-onset offending: Explorations in the concept of negative life events as turning points. *International Journal of Law, Crime and Justice, 43*(1), 1–16.

Farquhar, J. D., & Rowley, J. (2009). Convenience: A services perspective. *Marketing Theory, 9*(4), 425–438.

Fowler, T. (2014). BP's new tactic in oil spill claims: Go after the 'special master'. *The Wall Street Journal*. www.wsj.com, published January 27.

Freeh, L. J. (2013). *Independent external investigation of the deepwater horizon court supervised settlement program, report of special master Louis J. Freeh*. www.laed.uscourts.gov, published September 6.

Friedrichs, D. O. (2010). Integrated theories of white-collar crime. In F. T. Cullen & P. Wilcox (Eds.), *Encyclopedia of criminological theory* (Vol. 1, pp. 479–486). Sage.

Galvin, M. A. (2020). Gender and white-collar crime – Theoretical issues. *Criminal Justice Studies*, published online, https://doi.org/10.1080/1478601X.2020.1709954.

Galvin, B. M., Lange, D., & Ashforth, B. E. (2015). Narcissistic organizational identification: Seeing oneself as central to the organization's identity. *Academy of Management Review, 40*(2), 163–181.

Gamache, D. L., & McNamara, G. (2019). Responding to bad press: How CEO temporal focus influences the sensitivity to negative media coverage of acquisitions. *Academy of Management Journal, 62*(3), 918–943.

Goldstraw-White, J. (2012). *White-collar crime: Accounts of offending behavior*. Palgrave Macmillan.

Gormley, T. P. (2009). *Administrative and criminal penalties for illegal pollution under pollution control act*. International Law Office. https://www.internationallawoffice.com/Newsletters/Environment-Climate-Change/Norway/Arntzen-de-Besche-Advokatfirma-AS/Administrative-and-Criminal-Penalties-for-Illegal-Pollution-under-Pollution-Control-Act, publishes September 14.

Gottschalk, P. (2014). *Financial crime and knowledge workers – An empirical study of defense lawyers and white-collar criminals*. Palgrave Macmillan.

Gottschalk, P. (2017). Convenience in white-collar crime: Introducing a core concept. *Deviant Behavior, 38*(11), 605–619.

Gottschalk, P. (2019). *Convenience triangle in white-collar crime – Case studies of fraud examinations*. Edward Elgar Publishing.

Gottschalk, P., & Benson, M. L. (2020). The evolution of corporate accounts of scandals from exposure to investigation. *British Journal of Criminology, 60*, 949–969.

Greife, M. J., & Maume, M. O. (2020). Do companies pay the price for environmental crimes? Consequences of criminal penalties on corporate offenders. *Crime, Law and Social Change, 73*, 337–356.

Gulating. (2013, March 8). Court case LG-2011-60302 – LG-2012-23847, *Gulating lagmannsrett (Gulating court of appeals)*, Bergen, Norway.

Gulating. (2015, February 16). Court case 14-039291ASD-GULA/AVD1, *Gulating lagmannsrett (Gulating court of appeals)*, Bergen, Norway.

Hamilton, S., & Micklethwait, A. (2006). *Greed and corporate failure: The lessons from recent disasters*. Palgrave Macmillan.

Heltne, L., & Klevstrand, A. (2019). Her står Frederik Selvaags hytte ferdig – og ubrukelig. Fredrik Selvaag har fått nei fra Færder kommune til å ta i bruk sin nye Tjøme-hytte. Kommunen ber i stedet om en ny strandsone-søknad for hele hytta [Here stands Fredrik Selvaag's cabin finished – and useless. Fredrik Selvaag got no from Færder municipality to move into his new Tjøme cabin. The municipality asks instead for a new shoreline application for the complete cabin], daily Norwegian local newspaper *Tønsbergs Blad*. www.tb.no, published July 17.

Higgins, E. T. (1997). Beyond pleasure and pain. *American Psychologist, 52,* 1280–1300.

Holm, R. C. (2019). Hermine Midelfart risikerer å måtte rive deler av hytta på Tjøme [Hermine Midelfart risks having to tear down parts of the cabin on Tjøme], Norwegian public radio *NRK* Vestfold. www.nrk.no, published September 5.

Holm, R. C. (2020). To personer tiltalt for grov korrupsjon i hyttekommune [Two persons charged with serious corruption in cabin municipality], Norwegian public radio *NRK Vestfold*. www.nrk.no, published April 30.

Holmøy, E. (2017). Lederen av kontrollutvalget på Tjøme: -Betenkelig det som har forekommet [Leader of the control committee on Tjøme: -Worrysome what has taken place], local Norwegian daily newspaper *Tønsbergs Blad*, Wednesday, October 4, p. 6.

Holmøy, E. (2018). Byggesakslederen lot seg ikke stoppe av lover og regler – fjernet svaberg, lagde superanneks og grillhus med kjøkken og bad [The building case leader did not let himself be stopped by laws and rules – removed rocks, built annex and grill house with kitchen and bathroom], local Norwegian daily newspaper *Tønsbergs Blad*. www.tb.no, published December 20.

Holmøy, E. (2019). Denne terrassen skal bli mindre – må rive ulovlig del [This terrace shall become smaller – must remove illegal part], daily Norwegian local newspaper *Tønsbergs Blad*. www.tb.no, published December 27.

Holmøy, E. (2020). Ny anmeldelse på vei til Økokrim: -Vi har levert omfattende dokumentasjon [New report on its way to, Økokrim: -We have delivered comprehensive documentation], local daily newspaper *Tønsbergs Blad*. www.tb.no, published June 25.

Holth, F., & Winge, N. K. (2018, May 22). Bekymringer etter Tjøme-saken [Worries after the Tjøme case], Norwegian daily business newspaper *Dagens Næringsliv*, Tuesday, p. 20.

Huisman, W., & Erp, J. (2013). Opportunities for environmental crime. *British Journal of Criminology, 53,* 1178–1200.

Karim, K. E., & Siegel, P. H. (1998). A signal detection theory approach to analyzing the efficiency and effectiveness of auditing to detect management fraud. *Managerial Auditing Journal, 13*(6), 367–375.

King, M. (2020a). What makes a successful corporate investigator – An exploration of private investigators attributes. *Journal of Financial Crime*, published online https://doi.org/10.1108/JFC-02-2020-0019.
King, M. (2020b). Out of obscurity: The contemporary private investigator in Australia. *International Journal of Police Science and Management*, published online https://doi.org/10.1177/1461355720931887.
Klevstrand, A. (2018, Jule 20). Nedgraderer Tjøme-arkitekt [Downgrade Tjøme architect], Norwegian daily business newspaper *Dagens Næringsliv*, Wendesday, p. 17.
Klevstrand, A., & Randen, M. (2018, July 25). Suzuki importør bygget Tjøme-luksus på overtid: Risikerer å måtte rive hytta [Suzuki importer built Tjøme luxury on overtime: Risks having to tear down the cabin), Norwegian daily business newspaper *Dagens Næringsiv*, Wednesday, p. 1.
Klevstrand, A., & Solem, L. K. (2020, May 2). Økokrim undersøker hytteeiere, tiltaler arkitekt [Økokrim investigates cabin owners, charge architect], Norwegian daily business newspaper *Dagens Næringsliv*, Saturday, pp. 12–13.
Klevstrand, A., Solem, L.K., Dugstad, L. and Skaalmo, G. (2020). Tjøme-arkitekt Rune Breili tiltalt for grov korrupsjon – ga arkitekt-råd for 50 000 kr [Tjøme architect Rune Breili indicted for serious corruption – gave architect advice for 50,000 kroner], Norwegian daily business newspaper *Dagens Næringsliv*. www.dn.no, published April 30.
Krippendorff, K. (1980). *Content analysis: An introduction to its methodology*. Sage.
Krohn, M. D., & Eassey, J. M. (2014). Integrated theories of crime. In J. M. Miller (Ed.), *The encyclopedia of theoretical criminology* (pp. 458–463). Wiley.
Larssen, L. D. (2017). Tjøme kommune svidde av 200.000 kr på kommunikasjonsbyrå [Tjøme municipality spent 200,000 kroner on communication bureau], daily Norwegian local newspaper *Tønsbergs Blad*. www.tb.no, published December 19.
Lehman, D. W., Cooil, B., & Ramanujam, R. (2019). The effects of rule complexity on organizational noncompliance and remediation: Evidence from restaurant health inspections. *Journal of Management*, published online, pp. 1–33. https://doi.org/10.1177/0149206319842262.
Lien, A. O., & Aunemo, E. W. (2019). Hun får betalt av fleire som er dypt involvert i Tjøme-skandalen. Nå har hun samlet byggebransjen mot kommunen [She gets paid by many who are deeply involved in the Tjøme scandal. Now she has gathered the construction business against the

municipality], daily Norwegian local newspaper *Tønsbergs Blad*. www.tb.no, published May 24.

Lien, A. O., & Brevik, H. C. (2020). Frp-Anders forlot møte i protest: -Det ble rett og slett for drøyt [Right-wing Anders left the meeting in protest: -It simply became too much], daily Norwegian local newspaper *Tønsbergs Blad*. www.tb.no, published March 9.

Liska, A. E., Krohn, M. D., & Messner, S. F. (1989). Strategies and requisites for theoretical integration in the study of crime and deviance. In S. F. Messner, M. D. Krohn, & A. E. Liska (Eds.), *Theoretical integration in the study of deviance and crime* (pp. 1–20). State University of New York Press.

Logan, M. W., Morgan, M. A., Benson, M. L., & Cullen, F. T. (2019). Coping with imprisonment: Testing the special sensitivity hypothesis for white-collar offenders. *Justice Quarterly, 36*(2), 225–254.

Lynch, M. J. (2020). Green criminology and environmental crime: Criminology that matters in the age of global ecological collapse. *Journal of White Collar and Corporate Crime, 1*(1), 50–61.

Mai, H. T. X., & Olsen, S. O. (2016). Consumer participation in self-production: The role of control mechanisms, convenience orientation, and moral obligation. *Journal of Marketing Theory and Practice, 24*(2), 209–223.

Mann, K. (1985). *Defending white-collar crime: A game without rules*. Yale University Press.

Maslow, A. H. (1943). A theory of human motivation. *Psychological Review, 50*(4), 370–396.

McClelland, P. L., Liang, X., & Barker, V. L. (2010). CEO commitment to the status quo: Replication and extension using content analysis. *Journal of Management, 36*(5), 1251–1277.

Nordhordland. (2010, March 26). Court case 09-098460MED-NOHO, *Nordhordland tingrett (Nordhordland district court)*, Bergen, Norway.

Obodaru, O. (2017). Forgone, but not forgotten: Toward a theory of forgone professional identities. *Academy of Management Journal, 60*(2), 523–553.

Olaussen, M. (2018). Hvor mye koster det egentlig samfunnet at én privilegert mann bygger ulovlig i strandsonen? [How much does it really cost society that one single privileged person builds in the beach zone?], Norwegian daily local newspaper *Tønsbergs Blad*. www.tb.no, published May 15.

Patrucco, A. S., Luzzini, D., & Ronchi, S. (2017). Research perspectives on public procurement: Content analysis of 14 years of publications in the Journal of Public Procurement. *Journal of Public Procurement, 16*(2), 229–269.

Pedersen, N. (2017). "Professoren" sin kamp for å reinvaska seg ["The professor's" struggle to clean himself], public Norwegian broadcasting *NRK*. www.nrk.no, published August 19.
Piquero, N. L. (2012). The only thing we have to fear is fear itself: Investigating the relationship between fear of falling and white-collar crime. *Crime and Delinquency, 58*(3), 362–379.
Piquero, N. L. (2018). White-collar crime is crime: Victims hurt just the same. *Criminology & Public Policy, 17*(3), 595–600.
Pontell, H. N., Black, W. K., & Geis, G. (2014). Too big to fail, too powerful to jail? On the absence of criminal prosecutions after the 2008 financial meltdown. *Crime, Law and Social Change, 61*(1), 1–13.
Pratt, T. C., & Cullen, F. T. (2005). Assessing macro-level predictors and theories of crime: A meta-analysis. *Crime and Justice, 32,* 373–450.
Pratt, M. G., Kaplan, S., & Whittington, R. (2019). Editorial essay: The tumult over transparency: Decoupling transparency from replication in establishing trustworthy qualitative research. *Administrative Science Quarterly*, 1–19. https://doi.org/10.1177/0001839219887663.
Qiu, B., & Slezak, S. L. (2019). The equilibrium relationships between performance-based pay, performance, and the commission and detection of fraudulent misreporting. *The Accounting Review, 94*(2), 325–356.
Rodriguez, P., Uhlenbruck, K., & Eden, L. (2005). Government corruption and the entry strategies of multinationals. *Academy of Management Review, 30*(2), 383–396.
Saunders, M., Lewis, P., & Thornhill, A. (2007). *Research methods for business students* (5th ed.). Pearson Education.
Schnatterly, K., Gangloff, K. A., & Tuschke, A. (2018). CEO wrongdoing: A review of pressure, opportunity, and rationalization. *Journal of Management, 44*(6), 2405–2432.
Schneider, S. (2006). Privatizing economic crime enforcement: Exploring the role of private sector investigative agencies in combating money laundering. *Policing & Society, 16*(3), 285–312.
Skavhellen, Ø., & Lien, A. O. (2018). Tønsbergs Blad vant pris for dekningen av byggeskandalen på Tjøme [Tønsbergs Blad won prize for reporting the construction scandal], Norwegian local daily newspaper *Tønsbergs Blad*. www.tb.no, published May 2.
Stadler, W. A., Benson, M. L., & Cullen, E. T. (2013). Revisiting the special sensitivity hypothesis: The prison experience of white-collar inmates. *Justice Quarterly, 30*(6), 1090–1114.

Sundström, M., & Radon, A. (2015). Utilizing the concept of convenience as a business opportunity in emerging markets. *Organizations and Markets in Emerging Economies, 6*(2), 7–21.

Sutherland, E. H. (1939). White-collar criminality. *American Sociological Review, 5*(1), 1–12.

Sutherland, E. H. (1983). *White collar crime – The uncut version*. Yale University Press.

Sykes, G. M., & Matza, D. (1957). Techniques of neutralization: A theory of delinquency. *American Sociological Review, 22*(6), 664–670.

Thompson, R. (2017). With BP settlement claims winding down, Lafayette lawyer Parick Juneau turns attention to Takata air bag recall. *The Advocate*. www.theadvocate.com, published October 29.

Tilsynsrapport. (2020). Gryteskjær fritidsbebyggelse tilsynsrapport (Leisure Building Supervision Report). Færder kommune [Municipality of Færder], reference 19/49893, dated January 17, 4 pp.

Uribe, J., Sytch, M., & Kim, Y. H. (2020). When friends become foes: Collaboration as a catalyst for conflict. *Administrative Science Quarterly, 65*(3), 751–794.

Vestfold tingrett. (2020). *Case 17-145344TVI-VEFO/TO*, Vestfold District Court, verdict date September 17, sentence 25 pp., Tønsberg, Norway.

Wang, M. (2017). Informasjon eller tåkelegging: Det er bare to grunner til at kommunen ringer PR-rådgiver [Information or obfuscation: There are only two reasons why the municipality calls pr consultants], Norwegian daily local newspaper *Tønsbergs Blad*. www.tb.no, published December 20.

Wang, M. (2018). Teknologiriking vil bygge fritidsbolig på 400 kvadratmeter [Technology rich wants to build vacation home of 400 square meters], Norwegian local newspaper *Tønsbergs Blad*. www.tb.no, published September 23.

Wang, M. (2019). Mer trøbbel for hyttemillionær. Kommunen mener alle gamle tillatelser er ugyldige [More trouble for cabin millionaire: The municipality says that all old permits are invalid], Norwegian local newspaper *Tønsbergs Blad*. www.tb.no, published December 5.

Wells, J. B., Minor, K. I., Lambert, E. G., & Reeves, A. (2020). An exploratory study of possible correlates of individual whistleblowing propensity among sworn staff in a city jail. *Criminal Justice Policy Review*, 1–29, published online https://doi.org/10.1177/0887403420919478.

Wikstrom, P. O. H., Mann, R. P., & Hardie, B. (2018). Young people's differential vulnerability to criminogenic exposure: Bridging the gap between people- and place-oriented approaches in the study of crime causation. *European Journal of Criminology, 15*(1), 10–31.

Williams, J. W. (2005a). Reflections on the private versus public policing of economic crime. *British Journal of Criminology, 45,* 316–339.

Williams, J. W. (2005b). Governability matters: The private policing of economic crime and the challenge of democratic governance. *Policing & Society, 15*(2), 187–211.

Williams, J. W. (2008). The lessons of Enron: Media accounts, corporate crimes, and financial markets. *Theoretical Criminology, 12*(4), 471–499.

Williams, J. W. (2014). The private eyes of corporate culture: The forensic accounting and corporate investigation industry and the production of corporate financial security. In K. Walby & R. K. Lippert (Eds.), *Corporate security in the 21st century – Theory and practice in international perspective* (pp. 56–77). Palgrave Macmillan.

Wingerde, K., & Lord, N. (2020). The elusiveness of white-collar and corporate crime in a globalized economy. In M. L. Rorie (Ed.), *The handbook of white-collar crime* (chapter 29, pp. 469–483). Wiley.

Ytreberg, R. (2018, April 30). Kommunen stoppet Breili-prosjekt på Tjøme [The municipality stops Breili project on Tjøme], Norwegian daily business newspaper *Dagens Næringsliv*, Monday, pp. 4–5.

6

Attorney Defense Strategies

The competence of attorneys does matter for the outcome as illustrated by research in Norway, where Grendstad (2020) phrased the question "How much does it mean to have a better lawyer than the other party?" in an article:

> A survey of 800 Supreme Court rulings shows that if one party has a more experienced lawyer than the other party, the statistical chance of winning the case increases. While research shows that the facts of the case and the composition of the judges have an impact on the outcome of a case in court, it has been difficult to examine the importance of having an experienced lawyer.
>
> Over time, lawyers have increasingly become specialists, while judges remain generalists. In 2018, the Norwegian Supreme Court published their recommendations, in which the court provides the parties and lawyers with clear guidelines. The requirements are tightened for the lawyers. But does having an experienced lawyer in the Supreme Court make any difference?
>
> In an article in Justice, Olav Bjørnebekk, William Shaffer, Eric Waltenburg and I examined over 800 civil economic cases between a public and a private party in the Supreme Court. In addition to studying

P. Gottschalk, *Detecting and Investigating Environmental Crime*, https://doi.org/10.1007/978-3-030-74184-6_6

> the composition of judges, the financial issue in each case, and the outcome of the cases in district courts and courts of appeal, we examined whether the lawyers' experience in the Supreme Court had an impact on the outcome of the case… The survey showed that if one party has a more experienced lawyer than the other party, this party has an eight percent greater chance of winning the case.

In this chapter, Table 6.1 partly repeats Table 5.1 and lists wrongdoing by white-collar clients that their attorneys handle toward public authorities. All these issues came up in the last three years. The task for the attorneys is to prevent any reversal of any permits and to secure new permits to emerge.

Three Defense Strategies

Three themes are particularly noteworthy when distinguishing white-collar defense strategies from other defense strategies by attorneys. First, the role of white-collar attorneys is radically different from the typical criminal lawyer who defends a person suspected of and charged with street crime. For instance, the former spends far more time on each case, both in terms of workload and in terms of calendar time. Poverty and powerlessness tend to be the cause of street crime, while excessive power and greed are often the cause of white-collar crime.

This implies that a white-collar attorney works on fewer cases in parallel. The white-collar attorney gets a case much earlier and is far more likely to succeed in keeping wrongdoing charges away from the client. Second, information control is at the center of the white-collar attorney's work. The attorney is concerned with acquisition of crucial information and keeps damaging information out of the hands of public authorities such as municipalities, police investigators, and public prosecutors. A third theme centers on a major dilemma of these attorneys: how to defend vigorously the client without thereby becoming a party to the criminal act (Gottschalk, 2014).

In line with these themes, three specific strategies applied by white-collar attorneys are typical in defense of rich and mighty clients. First, a

Table 6.1 Identified white-collar clients and their cases of wrongdoing

	White-collar client	Attorney case
1	Ole Andresen	Denied swimming pool in Flatås road 109
2	Rune Breili	Architect involved in deviant cases
3	Arnfinn Hafsteen	Serious material competence shortage
4	Ole Borge	Decision to demolish a building
5	Arnfinn Hafsteen	Legality control of new building
6	Rune Breili	Suspected of corruption by the police
7	Werner Osther	Illegal blasting performed by approved company
8	Nanna Fløgstad Sohlberg	Summer house in Langestrand road 71 postponed
9	Gunnar Fredrik Selvaag	Summer house in Nautnes road 129 nonconforming
10	Hans Ole Helling	Demolition of tennis court
11	Carl Erik Krefting	Ottos road 46 liable for replacement
12	Turid Varner	Flekken road 64 with illegal terrace
13	Hans Petter Abrahamsen	Accusation of corruption by receiving bribes
14	Harald Dybsjord	Assessment cannot be reviewed afterwards
15	Sigmund Arnesen	Revocation of permission for Mostranda Camping
16	Harald Svendsen	Approval of own building applications
17	Rune Breili	Violation charges for Bukkeliodden 54
18	Kristina M. Wilhelmsen	General permit for Ødekjær road 43 reversed
19	Lars Hellik Strøm	Removal of terrace and splashing walls
20	Sigmund Arnesen	Decision on reversal for Mostranda Camping
21	Tore Jacobsen	Mo road 104 connection to Mostranda Camping
22	Øyvind Garner	Summer house moved five meters
23	Fridtjov Botvid Falck	Removal of stalls and exterior tiles
24	Celestine Clauson	No tennis court or swimming pool

(continued)

Table 6.1 (continued)

	White-collar client	Attorney case
25	Nanna Fløgstad Sohlberg	Postponed of cottage in Langestrand road 71
26	Harald Svendsen	Accusation of illegal construction in Sjøbodsand
27	77 construction firms	Construction industry lacking work
28	Rune Breili	Construction cases since 2010
29	Svein Homstvedt	Case of illegality in Flekken road 96
30	Scammer and scammed	Neholm road 67 fraud of 10 million kroner
31	Malling og Midelfart	Gryte Reef blown up and building without permit
32	Tom Ragnar Wilhelmsen	Lilleskag road 89, doubled cottage size
33	Anders C. Wilhelmsen	General permit for Ødekjær road 43 removed
34	Mette Landgraff	Violated property borders
35	Evy Reimers	Illegal boathouses in Dals road 88
36	Marianne and Ulf Stuge	Illegal terrace in Nord harbor 62

substance defense strategy is concerned with when and how an attorney decides to defend the client in a substantive way. Often, the substantive defense starts at a much earlier stage than in a street-crime case. Several of the attorneys listed in Table 5.1 in the previous chapter went early on into arguments with the municipality that all permits given were legally binding for the municipality.

Second, information control strategy is concerned with how the attorney determines what and how crucial information to control to make it difficult, and sometimes impossible, for public authorities to get the complete picture. Often, information control defense is able to keep secrets and to claim that pieces of information are irrelevant. In municipal control work, there are normally a number of information sources. Controlling and limiting some sources can ensure that municipal control procedures are never successful in solving the wrongdoing puzzle. Several of the attorneys listed in Table 5.1 went early on into the supply of documents in favor of their clients to the municipality, while they denied access to documents in disfavor of their clients.

Third, the symbolic defense strategy addresses all other means that the attorney can apply to divert attention away from legal issues. An example is to use the press or other media to portray the offender as a victim. This was particularly evident when attorney Vibeke Hein Bæra in defense of architect Rune Breili (line 6 in Tables 5.1 and 6.1) portrayed the architect as a victim. His architectural firm had almost collapsed and faced bankruptcy. Breili had bribed nobody in the municipality, but rather helped the head of building permits when that person needed an architect to design his new home (Klevstrand et al., 2020).

Substance Defense Strategy

The counterpart for an attorney in misconduct and crime cases is the authorities, who decide whether a suspect is required to demolish the summerhouse. To prevent a negative municipal decision, the attorney will try to convince the staff as well as the politicians in the municipality that the client has done nothing that justifies demolition. The strategic issue for the attorney is how to succeed stopping the public authority from advancing the inquiry from suspicion to decisions that harm the client. While very different from other crime cases, an attorney's active defense work often starts in the initial phases of detection, when there are only rumors of wrongdoing that may or may not be relevant for municipal interference (Gottschalk, 2014).

Initially, it does not matter for the attorney whether the client has actually done something wrong that deserves punishment, because rumors so often develop into accusations in the media. In the Tjøme case, the key whistleblower continued to feed the media with information about summerhouses that were not according to permits and landscapes that had changed completely. Investigative journalists got into boats and took pictures of the palaces. Both local media (e.g., Holmøy, 2017; Wang, 2019) and national media (e.g., Dugstad & Skaalmo, 2017; Klevstrand & Solem, 2020) published stories with summerhouse pictures and names of owners. Journalists called owners for comments, and the owners transferred the inquiries to their attorneys. In that way, attorneys typically involve themselves long before any potential sanctions are on the agenda in the town hall in the municipality of Tjøme.

If a white-collar attorney were to behave in white-collar cases similar to attorneys in street-crime cases, then the attorney would simply wait for the arrival of evidence against the client. Then, the attorney would react to the evidence and the potential sanction. The street-crime attorney is mainly reactive, while the white-collar attorney is proactive. The street-crime attorney waits for formal charges against the client and then makes up an opinion what to do next. Typically, the attorney argues that the evidence is not sufficient to prove wrongdoing or guilt because presented evidence does not document in an adequate and convincing manner that the client has done something wrong that requires sanction. The attorney may argue that the evidence only proves outdated misconduct that cannot lead to sanctions. Uncertainty and doubt should benefit the client, every attorney will argue. It is every attorney's job to interpret laws and regulations in a subjective way, which should lead to a not-guilty conclusion for the client. This is the substance defense strategy. The attorney contributes legal and factual substance to a case handled by a public authority such as a municipality.

White-collar attorneys also contribute substance defense like street-crime lawyers, but there are several differences, related both to points in time and to magnitude of contribution. A white-collar substance defense starts much earlier in the public agency process, and it may last much longer into all kinds of appeals and retrials. The magnitude of contribution relates to the level of detail and the scope of evidence. Often, financial wrongdoing finds its documentation in hidden details, such as a single bank transaction, a single invoice, or a single procurement of building material. The scope of evidence is often such that a lot of redundant and irrelevant material overflows the case to make sure that the parties overlook nothing of potential relevance. Instead of waiting for a potential sanction from the municipality, such as demolition of parts of the palace, or waiting for the criminal case to open in court, the attorney starts substance defense as early as possible potentially to stop the procedure and the client from ever appearing in a municipal case or a court case. It is more the rule than exception that the attorney, in fear of future consequences, is actively on the case already when there are merely suspicion and rumors circulating. More rumors and accusations in the public domain can in fact be helpful to the attorney, as the attorney may argue that the case needs a stronger focus.

For example, in a fraud case in Norway, police investigators accused a family of insider trading in a company where they were involved. National media heavily covered the case. The family was a major owner of a large Norwegian manufacturing company, but there were other shareholders as well. Initially, in the case, the police investigated the father and the mother as well as the daughter and son. A defense lawyer for the daughter successfully helped to keep her out of the case, arguing that the police should stay more focused. Despite the fact that the police found an SMS on her phone saying 'papa,' which obviously meant her father, police detectives accepted her explanation that she called many older male persons, such as her uncle, for 'papa,' and the prosecution thus dismissed her case and never prosecuted her in court. The same thing happened to the mother. Only the father and son faced prosecution, and they received convictions to long prison sentences in court.

Similarly, the rich and mighty on the island of Tjøme had attorneys listed in Table 6.1 who argued that their clients did not know how architects, builders, plumbers, carpenters, electricians, landscapers, and other professionals applied for and obtained various permits from the municipality. The attorneys argued that the municipality and potentially the police should focus their cases on those who had submitted misleading applications, and who had deviated in practice from their own applications when doing jobs on elite members' properties on the island. The people who had done the job of creating tennis courts, swimming pools, annexes, park facilities, docks, boathouses, and other surprising elements should be addressed, not to the clients who had hired the people to do the job. It was all about focusing on the relevant offenders, the attorneys argued. This was substance defense in terms of influencing the definition of facts.

Weisburd et al. (1991: 99) formulated in the Yale Series after Mann (1985) how the white-collar defense lawyer is involved at a very early stage:

> From the time there is even a hint that a possible white-collar crime is under investigation by legal authorities, individuals suspected of involvement often begin to retain attorneys and to prepare to defend themselves. Early legal strategies may include negotiations with the agencies involved, the seeking of civil or out-of-court resolution of the case, and the trading of information in return for favorable treatment from the prosecutor's office. Other strategies include defense efforts to limit the scope of the information sought through subpoenaed documents and to curtail the information obtained by the government through search warrants and electronic surveillance.

Bjørn Stordrange, a well-known white-collar lawyer in Norway, defended Acta entrepreneur Fred Anton Ingebrigtsen, who faced accusations of insider trading in the Acta stock by helping family members make illegal profits. Early on, Stordrange expressed in public his frustration with several delays in police investigations (Haakaas, 2009: 2):

> When the charge was out last summer, we were told that the investigation would be completed last fall. The time limit was changed to Easter and then again to October this year. Now we are told that it might be completed by Christmas.

Stordrange's many appearances in the media indicated both active substance defense in the Acta case and symbolic defense, which is a topic later in this chapter. Stordrange was extremely proactive in terms of an organized crime suspicion, where the police wanted to use the Norwegian mafia rule on the case. Stordrange succeeded in convincing the police to drop the organized crime charge against Ingebrigtsen and his crime associates.

Similarly, the rich and mighty on the island of Tjøme, who had all used the same architect to make their dreams come true, distance themselves from any pattern suggested in the media. While they use the neutralization technique of normality, where everyone else did what they did, they also claim that they did not notice what everyone else did other than what the architects and others told them.

More importantly in their substance defense, attorneys identified a number of mistakes in procedures in the municipality when handling client cases. There were delays, role conflicts between bureaucrats and politicians, and rule complexity that were impossible to understand for the nonprofessional (Lehman et al., 2019). Attorney Frode S. Jørgensen argued for his client Harald Dybsjord that the municipality simply did not have any relevant legal basis for reviewing previous approvals.

In Norway, white-collar attorneys are typically trying to influence police investigations at Økokrim. Økokrim is the main source of specialist skills for the police and prosecution authorities in their combat against financial and environmental crime (www.okokrim.no). Økokrim is similar to the Serious Fraud Office (SFO) in the United Kingdom. The SFO is an independent government department, operating under the superintendence of the Attorney General. Its purpose is to protect society by investigating and, if appropriate, prosecute those who commit serious or complex fraud, bribery, and corruption and pursuing them and others to recover the proceeds of their crime. For example, on March 11, 2013, the SFO charged three men in a Ponzi-style scheme, and the men appeared at City of London Magistrates Court charged with conspiracy to defraud investors in an alleged investment fraud related to electrical contacts in the hotel sector (www.sfo.gov.uk).

Crime in the legal sense is about the facts and the laws. If the facts do not match the laws, then Økokrim is reluctant to investigate the case. Very often, also in the environmental crime cases on the island of Tjøme, both facts and laws are unclear. The lack of clarity opens up for skilled attorneys to argue that Økokrim should stay away from the cases. The attorneys present client cases early on in such a way that it seems that the cases do not match the statutes. For example, many Norwegian laws have a limitation period where offenses become obsolete. Attorneys thus argue for a date of wrongdoing that causes the obsolescence clause to apply.

Information Control Strategy

Attorneys tend to take charge of information control at an early stage. Instead of being at the receiving end of documents from public authorities, the attorney is in a position where the attorney can monitor the flow of information. Of particular interest to the attorney is crucial information that can harm the client's case. The flow of harmful facts, insights, and knowledge of causes and effects that might become legal evidence with the public authorities is problematic for the client, and therefore, the attorney attempts to stop the flow.

An element of information control is control over various parties' explanations. The attorney attempts to keep parties in wrongdoing in line with each other to avoid a blame game between people who had different roles. If the summerhouse owner starts to blame the architect, the architect starts to blame the builder, the builder starts to blame the landscaper, and so on, then information will surface that typically harm all of them. This is in line with the prisoner's dilemma, in which individuals acting in their own self-interests do not produce the optimal outcome. Parties choose to protect themselves at the expense of other participants. As a result, participants find themselves in a worse state than if they had cooperated with each other.

This is exactly what happened in a different shoreline case in Norway, where the rich and mighty Hanne Madsen received a verdict of 45 days in jail by the court. She had bought a property behind her summerhouse, which was within the one-hundred-meters belt as well. On the acquired property, she built an illegal tennis court (DN, 2019). She presented herself as a victim of law violations and blamed the carpenter who eventually received a prison sentence as well. Agder (2019: 20) court of appeals wrote in its verdict:

> There is a strong development pressure in the beach zone and people invest very high amounts to gain access to this limited good. Effective enforcement of established rules and guidelines is therefore necessary, and material and serious violations must result in tangible reactions. Often, purely economic reactions will have less deterrent effect.

Strategic substance defense is not necessarily the first defense strategy applied by the attorney in a white-collar case. The defense lawyer's very first goal can be to prevent public authorities from obtaining information and evidence that is harmful to the client and prevent information from application by the criminal justice systems and other authorities to define and justify a formal charge or sanction.

At this stage, not laws and verdicts are of concern to the attorney. All the attorney is worried about is the flow of information that may transform into potential evidence for municipal sanctions and police investigations (Holmøy, 2020a). The attorney's job is all about preventing public authorities—especially the criminal justice system—from acquiring evidence and making it difficult or even impossible for them to understand pieces of information that they have obtained. It is all about stopping the inquiry at an early stage so that the case is closed. This is the defense lawyer's information control strategy (Gottschalk, 2014).

Information control implies that documents can remain confidential and hidden, and that clients and witnesses do not talk to officials in public positions. It may also imply that individuals enjoy protection from the press, so that only the lawyer makes statements about the case. Thereby, avoidance of leaking unfavorable information is more likely. Summerhouse owners in Table 6.1 paid their attorneys to represent themselves in different contexts including the media. Attorneys then emphasized information that was in favor of their clients and denied negative accusations and rumors about their clients. As is evident from the table, one attorney, in particular, Anne Tellefsen at local law firm Haraldsen Bydal Lie, represented many rich and mighty wrongdoers. She became an expert in handling different authorities and the media for her clients.

If Tjøme cases would escalate, then some clients are likely to replace local attorneys with nationally reputable attorneys. However, as long as the idea is to portray themselves as hunted innocent individuals suffering from the local municipality's actions, then it may seem smart to pay local attorneys. Clients do not want their cases to escalate. Rather, they would like their cases to close so that the cases move into public memory loss. Therefore, an information control strategy is to tell as little as possible to investigative journalists when they call.

Information control strategy finds application ahead of substance defense strategy whenever a white-collar offender first faces accusations in the media, which then cause initial signal alertness among public authorities. If the lawyer is successful in strategic information control, then raw material for legal argumentation remains hidden from public attention and use. The case for sanctions or prosecution deteriorates because important pieces of information not known to the municipality are missing. If an inquiry represents a puzzle, where all pieces have to be in place to see the picture, then both missing pieces and ill-placed pieces will make it difficult to perceive, understand, and interpret the fragmented picture. Inspectors from the municipality may find themselves with a case that is impossible to solve and thus, decide to close it.

If a public agency is aware of information that they so far have not been able to collect, the attorney may argue that the requested information is difficult to retrieve and irrelevant for the case. The attorney may argue that the information is confidential, that it is out of date, or that it links to other problematic information. If the public agency already has gathered the information, the attorney may argue that the information does not apply to the specific case because authorities have collected it in an unethical or inadmissible way, such as pirate copying by other citizens, by trespassing to take pictures, or by violation of data protection laws.

These arguments when performing information control have the purpose of influencing the counterpart, either by convincing the public agency it is not a good idea to press for information or press charges or by obtaining a court ruling stating that information should not be made available or should even be returned to the client or the client's attorney. The attorney communicates procedure rules that support information control to the agency. For example, the attorney may argue that the law prohibits the search for or collection of specific documents. An agency may argue that the law allows it. Nevertheless, public officials must consider whether it is worth the fight with the attorney at this stage.

Strategic information control implies stopping or limiting the flow of information from the client to a public agency, by preventing the public agency access to informants who deny participation in interrogations, by preventing access to other information sources, and by requesting the return of documents not used in the agency (Mann, 1985: 7):

> The defense attorney's aim is to instruct the client or third party holding inculpatory information how to refrain from disclosing it to the government and, if necessary, to persuade or force him to refrain.

Inculpatory, also called incriminating, information is that which shows that a person has been involved in a criminal act. It is self-revealing information by the suspect. Inculpatory information finds application in the criminal justice system as evidence to prove guilt. Exculpatory information helps prove that a person has not been involved in a criminal act. In the information control strategy, the attorney attempts to stimulate the flow of exculpatory information and prevent the flow of inculpatory information.

An attorney's active information control strategy can remain secret to other parties in the case, including the client. Success is often dependent upon the lack of awareness among other parties, including the press.

When an attorney advises a client not to answer certain questions in the next interview and instead answer that he does not know or that he will have to check facts first, the client is subject to information control. The attorney is applying the information control strategy via the client. The public agency knows nothing about it. One result might be that the inquiry suffers from delay and potential termination.

For example, attorney Eilert Eilertsen for client Nanna Fløgstad Sohlberg attempted to delay case handling in the municipality by sending a letter in June 2018:

> The case appears to be very comprehensive and complex. For this reason, I need to have access to some documents in the municipality's possession before I can properly protect the family's interests.

One reason to delay the case might be to make it obsolete after a number of years. Many statutes in Norwegian law define offenses as outdated after some years. Even if a law violation is not yet outdated, an agency will be reluctant to look into a case that soon will become outdated.

A former police officer, now academic at a university in the United Kingdom, expressed the following opinion about white-collar solicitors in a personal e-mail communication:

> As an ex-police officer, I anecdotally know that solicitors lie and use all forms of diabolical half-truths to get clients off. They are entrepreneurial in their use of knowledge and of systems to get results. Similar to detectives as entrepreneurs, they are continually working, lurking, and getting results.

Information control does not only occur in white-collar cases. In many other cases, the attorney works to exclude pieces of evidence from agency access and application. The attorney may argue that acquired information in the agency is a result of an illegal manner of information collection, or that information is misleading or irrelevant to the case. What makes white-collar cases so special is that strategic information control is of key importance—it is sometimes the most important activity—successfully to work for the client's best interests. In other kinds of cases, information control is mainly a tactical maneuver to detract attention or delay the case temporarily.

The information control strategy finds support in the attorney-client privilege as well as the work-product privilege. While the attorney-client privilege shields any information communicated to an attorney, the work-product privilege protects information that represents preparation for litigation (Oh, 2004).

Strategic information control is concerned with the flow of damaging information about the client. An attorney will attempt to prevent a public agency from exploring and exploiting various sources of information collection. Strategic information control implies taking control over information sources that are most likely to receive requests from a public agency. A public agency has many information sources when they inquire into a case, and a defense attorney can influence these sources to a varying extent.

Symbolic Defense Strategy

A symbol is an object or a phrase that represents an idea, a belief, or an action. Symbol defense is the application of symbols that can benefit the client. An example is the symbol of 20 million Norwegian kroner (about 2 million US dollars) that attorney Anne Tellefsen applied to benefit her client Celestine Clauson. Tellefsen claimed that the amount represented what Clauson had paid for expansion rights when she bought the summerhouse (Holmøy, 2020a). Clauson paid 30 million Norwegian kroner for the summerhouse, but it was only worth 10 million without the expansion rights. Clauson's attorney claimed that the seller had told Clauson that the municipality had granted expansion rights. The claimed expansion approvals included a swimming pool, a tennis court, a larger basement, as well as changes to the façade of the building. When the whistleblower informed politicians in the municipality that approvals were not according to rules and regulations, the councilor recommended to the politicians to determine permits and approvals as illegitimate and not valid. The reaction from Tellefsen was to send a long letter to the politicians telling that the municipality would be liable for Clauson's loss of NOK 20 million kroner. One of the politicians expressed in the media that she perceived the letter as a threat (Holmøy, 2020a). The threat was symbolic, because the municipality said it is never involved in the private transaction of real estate between a seller and a buyer.

Attorney Henning Harborg on behalf of his client Carl Erik Krefting also applied a symbolic defense of threat. The attorney claimed in a letter to the municipality that the municipality would be liable to compensate his client completely for any economic loss that could occur because of the review of building matters by Krefting (Blix, 2018). Harborg was also attorney for the widow Turid Varner, but she decided to tear down the terrace and the park facility that blocked the crawl for people, instead of continuing the quarrel with the municipality.

Symbolic defense is concerned with activities that represent defense, but the activities in themselves are no defense. Symbols take the form of words, sounds, gestures, or visual images. When attorney Vibeke Bæra asked for a picture of herself with her client Rune Breili, the visual image published in the local newspaper signaled a hard-working architect who had done nothing wrong.

Symbolic defense is an alternative and supplement to substance defense. Substance and symbolic defense are different arenas where the white-collar attorney can work actively to try to make the public officials close the case. The purpose of symbolic defense is to communicate information and pseudo-legal opinions by means of symbols. Examples of attorney opinions are concerns about unacceptable delays in public administration, low quality work by public officials, or other issues related to work at public agencies. Several of the attorneys listed in the table for Tjøme applied this symbolic defense strategy. For example, attorney Stig L. Bech at law firm BAHR wrote a five-page letter to the municipality pointing at all the weaknesses in the local administration. Since the municipality is required to reply to questions phrased in the letter, they had to spend time on their reply rather than on doing casework, which caused a further deterioration in the deadline statistics for the administration.

Complaining about delays in inquiries and administrative work is not substance defense, as the complaint is not expressing a meaning about the wrongdoing and possible sanctioning and punishment. Complaining is a symbolic defense, where the goal is to mobilize sympathy for the white-collar client and to reduce the chances of success for the counterpart.

White-collar attorneys vary in the extent to which they find media coverage a suitable option for symbolic defense. Ideally, each attorney should only have what is best for the client in mind. However, some famous attorneys may have their own media agenda, more or less independent of the current clients they are supposed to help and assist. Irrespective of client and/or attorney agenda, the attorney must be prepared to respond to media inquiries. The client may want to present the story in the press because he or she may believe that press coverage will have a positive impact on personal image as well as proceedings in the public agency and in the criminal justice system. Readers may perceive public agency statements in the press as provocation against the client, who wants to tell a different story.

Two attorneys in Table 5.1 represented themselves, but they presented themselves as attorneys for clients that happened to be themselves. Attorney Ole Andresen complained about the reversal of decision in the municipality. He wanted to have a swimming pool and build an annex, but the municipality had withdrawn the permit. Attorney Ole Borge also represented himself. Tjøme municipality had ordered the removal of his suspension bridge and steel bars. He complained and tried to compare his case with other cases where approvals were still valid.

While the administration in the municipality handled complaints and reviewed permits and withdrawn permits, several attorneys skipped contacting the administration and sent mail to politicians directly instead. Typically, few days before a political body was to meet, the attorney sent a letter as an email attachment to each member of the political body. In that way, the attorney hoped to short-circuit the process in the administration. Tellefsen's letter on behalf of Clauson is an example. Another example is the rich heir Gunnar Fredrik Selvaag who wrote a personal letter to a political committee in the municipality with the help of his attorney Mari Helen Gulsvik.

Attorney Gulsvik at law firm Selmer also helped her client Selvaag use his summer mansion in Nautnes road 129 despite the refusal from the municipality. Heltne and Klevstrand (2019) reported next to a picture of the mansion:

> Here stands Fredrik Selvaag's cabin finished – and useless. Fredrik Selvaag got no from Færder municipality to move into his new Tjøme cabin. The municipality asks instead for a new shoreline application for the completed cabin.

However, attorney Gulsvik sent on July 9, 2019, an application to the municipality for a temporary permit to use the cabin. Carina M. Iversen-Nordby at Færder municipality granted such a permit on August 19, 2019. The temporary permit could enable Selvaag to get a permanent permit by arguing that a permit had already been granted. One of the two summerhouse owners that Økokrim investigated was Fredrik Selvaag, who had inherited a contracting company for apartment buildings, as described in the next chapter.

A frequent theme in the symbolic defense by attorneys was tennis courts and swimming pools. Some cases concerned the denial of permits to build tennis courts and swimming pools. Other cases concerned removal of tennis courts and swimming pools that had not received appropriate permits. Attorney Christian H.P. Reusch was engaged in a former case for summerhouse owner Kristina M. Wilhelmsen, while attorney Nils Hansson was engaged in a latter case for summerhouse owner Hans Ole Helling.

A common approach in symbolic defense is to portray wrongdoing as completely insignificant while the reaction by the public agency is abuse of power. An example is attorney Martin Støren for his client Lars Hellik Strøm. Støren wrote in his letter "the municipality may refrain from prosecuting violations if conditions are of minor importance."

Martin Støren is one of the symbolic defense lawyers in Table 5.1 representing three clients on the island of Tjøme. Støren is an attorney at law firm Schjødt in Oslo. He wrote a letter to the municipality of Færder on behalf of the owners of Mostranda Camping on May 28, 2020. The letter claimed that the municipality had caused substantial loss to the owners of Mostranda Camping. The letter is a typical example of attribution of blame (Eberly et al., 2011; Lee & Robinson, 2000; Resodihardjo et al., 2015; Xie & Keh, 2016) and shaming (Braithwaite, 1989). Attorney Cicilie Sørum Gregoire in the municipality replied June 16, 2020:

> As far as the municipality can see, there will in any case not be a sufficient causal connection between regulation costs and any liability-related action from the municipality. The constructions applied for would in any case not be able to be built without a new planning basis or exemptions. On the basis of the above, claims for compensation against the municipality in this case will be disputed. It is not documented that the conditions for compensation can be met.

The symbolic defense by attorney Anne Tellefsen on behalf of her client Celestine Clauson was interesting as the summerhouse case evolved in the media. When an investigative journalist interviewed the seller, that is the previous owner, Stein Kittelsen, he denied ever to have applied for permits to have a swimming pool and tennis court (Holmøy, 2020a: 8):

> We sold it, as it is, not with any permits at all. We think the property is worth around 30-35 million kroner, says Kittelsen.

When Celestine Clauson in 2015 bought the summerhouse in Nesholm road 67 on the island of Tjøme from the family Kittelsen, who had owned the place for generations, she was only 18 years old. Her father, Conrad Clauson, was the real owner. He had made a fortune as a broker when he worked for Credit Suisse First Boston in London, where he exploited bank clients for his personal benefit (Ahuja, 2005):

> Credit Suisse First Boston has started legal proceedings against a former member of staff after he failed to repay a $750,000 loan. CSFB has launched the claim, which was filed in the UK high court in February, against Conrad Clauson, a Norwegian who worked as a private client broker at Donaldson, Lufkin & Jenrette, which became part of Credit Suisse in August 2000.

Clauson had placed his fortune in a trust in the tax haven Virgin Island (Berglihn et al., 2007). In the Tjøme case, he first paid architect Breili successfully to apply for permits in the municipality. Then, he paid attorney Tellefsen to fight for the illegal permits by symbolic defense (Holmøy, 2020a).

Another aspect of attorney Anne Tellefsen's symbolic defense for her client Celestine Clauson, who had inherited wealth from her father Conrad Clauson (Ahuja, 2005; Berglihn et al., 2007; Holmøy, 2020a), was to engage in a quarrel with Færder municipality regarding her lawyer bill. Tellefsen claimed that the municipality had to pay for the time she spent when she was present during the review visit of municipal officials on Clauson's potential crime scene. Tellefsen complained all the way to the ministry that her millionaire client should have her legal expenses covered. The expenses amounted to the equivalent of USD 18,000. The answer from the ministry arrived in October 2020 (Holmøy, 2020b):

> On behalf of her client Celestine Clauson, Tellefsen has demanded 180,000 kroner from Færder municipality to cover her legal fees. She gets 68,750 kroner.

The symbolism of the requirement to cover legal fees rests on the likely fact that Tellefsen's struggle to get some money refunded probably was far more expensive than the actual refund demanded or received. Tellefsen had spent more than a year to appeal the request for refund via a number of authorities to the ministry. The symbolism effect could be that the municipality would get so exhausted from Tellefsen's many initiatives that they would be more reluctant to chase her clients who have illegal piers, extensions, lawns, terraces, and other facilities.

Bibliography

Agder. (2019, September 16). *Court case 18-178766AST-ALAG*, Agder lagmannsrett (Agder court of appeals), Agder, Norway, 22 pp.

Ahuja, V. (2005). CSFB sues former broker over $750,000 loan. *Financial News London*. www.fnlondon.com, published August 8.

Berglihn, H., Gjernes, K., & Nordahl, B. O. (2007). Metoderapport: Skjult i skatteparadis (Method Report: Hidden in Tax Haven), daily Norwegian business newspaper *Dagens Næringsliv*. www.dn.no, published January 18.

Blix, N. T. (2018). Økokrim får fortsatt mange Tjøme-saker og regner med at etterforskningen strekker seg langt ut i 2019 [Økokrim still receives many Tjøme cases and expect the investigation to last far into 2019], daily local Norwegian newspaper *Tønsbergs Blad*. www.tb.no, published November 9.

Braithwaite, J. (1989). *Crime, shame and reintegration*. Cambridge University Press.

DN. (2019, September 21). Ny dom kan skremme pengesterke hytteeiere fra ulovlig bygging – Ny dom gir god hjelp til kommuner som vil stoppe ulovlige byggearbeider i strandsonen [New court sentence might scare money-rich cottage owners from illegal construction building – New court sentence provides good help to municipalities who want to stop illegal construction work in the beach zone], daily Norwegian business newspaper *Dagens Næringsliv*, Saturday, p. 2.

Dugstad, L., & Skaalmo, G. (2017). Slik fikk han de rikes hytter godkjent på Tjøme [This is how he got the cabins of the rich approved on Tjøme], daily Norwegian business newspaper *Dagens Næringsliv*. www.dn.no, published September 29.

Eberly, M. B., Holley, E. C., Johnson, M. D., & Mitchell, T. R. (2011). Beyond internal and external: A dyadic theory of relational attributions. *Academy of Management Review, 36*(4), 731–753.

Gottschalk, P. (2014). *Financial crime and knowledge workers – An empirical study of defense lawyers and white-collar criminals*. Palgrave Macmillan.

Grendstad, G. (2020). Hvor mye betyr det å ha en bedre advokat enn motparten? [How much does it mean to have a better lawyer than the other party?], web-based legal newspaper *Rett24*. www.rett24.no, published October 20.

Haakaas, E. (2009, November 27). Acta-saken kan smuldre bort [The Acta case can mold away], daily Norwegian newspaper *Aftenposten*, Business section, p. 2.

Heltne, L., & Klevstrand, A. (2019). Her står Frederik Selvaags hytte ferdig – og ubrukelig. Fredrik Selvaag har fått nei fra Færder kommune til å ta i bruk sin nye Tjøme-hytte. Kommunen ber i stedet om en ny strandsone-søknad for hele hytta [Here stands Fredrik Selvaag's cabin finished – and useless. Fredrik Selvaag got no from Færder municipality to move into his new Tjøme cabin. The municipality asks instead for a new shoreline application for the complete cabin], daily Norwegian local newspaper *Tønsbergs Blad*. www.tb.no, published July 17.

Holmøy, E. (2017, October 4). Lederen av kontrollutvalget på Tjøme: -Betenkelig det som har forekommet [Leader of the control committee on Tjøme: -Worrysome what has taken place], local Norwegian daily newspaper *Tønsbergs Blad*, Wednesday, p. 6.

Holmøy, E. (2020a). Ny anmeldelse på vei til Økokrim: -Vi har levert omfattende dokumentasjon [New report on its way to, Økokrim: -We have delivered comprehensive documentation], local daily newspaper *Tønsbergs Blad*. www.tb.no, published June 25.

Holmøy, E. (2020b). Advokat Anne Tellefsen klaget helt opp til departementet for at hennes millionærklient skal få dekket utgiftene sine. Nå har svaret kommet [Lawyer Anne Tellefsen complained all the way to the ministry that her millionaire client should have her expenses covered. Now the answer has come], local daily newspaper *Tønsbergs Blad*. www.tb.no, published October 16.

Klevstrand, A., & Solem, L. K. (2020, May 2). Økokrim undersøker hytteeiere, tiltaler arkitekt [Økokrim investigates cabin owners, charge architect], Norwegian daily business newspaper *Dagens Næringsliv*, Saturday, pp. 12–13.

Klevstrand, A., Solem, L. K., Dugstad, L., & Skaalmo, G. (2020). Tjøme-arkitekt Rune Breili tiltalt for grov korrupsjon – ga arkitekt-råd for 50 000 kr [Tjøme architect Rune Breili indicted for serious corruption – gave architect advice for 50,000 kroner], Norwegian daily business newspaper *Dagens Næringsliv*. www.dn.no, published April 30.

Lee, F., & Robinson, R. J. (2000). An attributional analysis of social accounts: Implications of playing the blame game. *Journal of Applied Social Psychology, 30*(9), 1853–1879.

Lehman, D. W., Cooil, B., & Ramanujam, R. (2019). The effects of rule complexity on organizational noncompliance and remediation: Evidence from restaurant health inspections. *Journal of Management*, published online, pp. 1–33. https://doi.org/10.1177/0149206319842262.

Mann, K. (1985). *Defending white-collar crime: A game without rules*. Yale University Press.

Oh, J. J. (2004, March). How (un)ethical are you? Letters to the editor. *Harvard Business Review*, 122.

Resodihardjo, S. L., Carroll, B. J., Eijk, C. J. A., & Maris, S. (2015). Why traditional responses to blame games fail: The importance of context, rituals, and sub-blame games in the face of raves gone wrong. *Public Administration, 94*(2), 350–363.

Wang, M. (2019). Mer trøbbel for hyttemillionær. Kommunen mener alle gamle tillatelser er ugyldige [More trouble for cabin millionaire: The municipality says that all old permits are invalid]. Norwegian local newspaper *Tønsbergs Blad*. www.tb.no, published December 5.

Weisburd, D., Wheeler, S., Waring, E., & Bode, N. (1991). *Crimes oft he middle classes*. Yale University Press.

Xie, Y., & Keh, H. T. (2016). Taming the blame game: Using promotion programs to counter product-Hhrm crises. *Journal of Advertising, 45*(2), 211–226.

7
Tjøme Investigation Start

The local police was reluctant to investigate rumors of environmental crime on Tjøme Island. Two whistleblowers had detected a number of summerhouses on the shoreline that were not in accordance with municipal permits that they retrieved from municipal archives. The whistleblowers Amundsen and Ravndal informed politicians in the municipality as well as investigative journalists in the local newspaper 'Tønsbergs Blad' (Schjøll, 2020) and the national business newspaper 'Dagens Næringsliv' (Klevstrand & Espeland, 2020). Both newspapers sent photographers out in boats to take pictures of vacation homes on the island that the Amundsen and Ravndal had identified. Local residents commented on the news reports by asking for municipal intervention to stop the wrongdoing in the beach zone.

Local police had already received a formal report about rich people on the Gryte Reef from the municipality. Rather than initiating an investigation, local police closed the case. Therefore, the two whistleblowers contacted the national police agency responsible for investigation and prosecution of financial an environmental crime ('Økokrim'). The following chapters describe how the police started their investigations in 2018 that resulted in scheduled court hearings in 2021.

P. Gottschalk, *Detecting and Investigating Environmental Crime*,
https://doi.org/10.1007/978-3-030-74184-6_7

Tjøme municipality merged with neighboring municipality Nøtterøy in 2018. The name of the new municipality is Færder. While the potential crime cases remain referred to as the Tjøme cases or the overall Tjøme case here in this book, the handling of the matter continued in 2018 in Færder municipality. Tjøme municipality merged with Nøtterøy municipality into Færder municipality on January 1, 2018. Færder is a municipality in Vestfold and Telemark County. The administrative center of the municipality is the village of Borgheim. The 100-square-kilometer (39 square miles) municipality is the 380th largest by area out of the 422 municipalities in Norway.

The local police in Vestfold and Telemark County are located in the city of Tønsberg, the city of Larvik, and some more places in the county. The local police have a group of detectives assigned to investigation of financial crime, including environmental crime. However, as will become evident from the case presentation of the issues on Tjøme Island, the crime scene was too complicated for local detectives to handle. The crime scene covered several decades, several shoreline parts on the island, several employees in the municipality, several summerhouse owners and their attorneys, and an architectural firm. The local detective function lacked competence rather than capacity, so local police was reluctant to get involved in the investigation. While the initial reports about wrongdoing on the island of Tjøme arrived at local police stations, and the initial public awareness about wrongdoing was created in local media, the Tjøme case developed into a national matter at Økokrim (Norwegian authority for investigation and prosecution of economic and environmental crime) and in national media. Like local police, Økokrim as well was initially reluctant to look into the wrongdoing on the island. They asked the municipality formally to report cases to the police, where the municipality should formulate their suspicion of financial and environmental crime against selected actors on the island. However, the municipality was reluctant to express suspicions, accusations, and allegations against named individuals. The municipality argued that local police and Økokrim could investigate independent of formal reports with crime allegations. Færder municipality confirmed that they were willing to help the police with documentation and access to whatever information that was needed in the investigation, but the municipality did not want to accuse anyone. The municipality ended up reporting formally some of the summerhouse owners to Økokrim, as will become evident from the following case description. The initial passivity of the police and the municipality was compensated by the whistleblowers as informants and by local and national media uncovering new scandals that public officials eventually had to report and investigate.

Fraud Examiners from BDO

Before the police involved itself in the Tjøme case, the municipality hired fraud examiners from global auditing firm BDO (2017) to examine accusations of wrongdoing in the building permits department. The BDO examiners concluded that they could find no misconduct or crime. The mayor at that time was Bente Bjerke, and the councilor was Christine Norum (Holmøy, 2018). Dag Lars Erichsrud was the leader of the control committee in the municipality. After the release of the BDO report, the municipality spent 200,000 Norwegian kroner in a public relations effort to make inhabitants think that the case was closed (Larssen, 2017). However, inhabitants did not think so. Therefore, the control committee wanted to hire another global auditing firm, KPMG, for a new fraud examination, but the municipal council refused in October 2017 to provide funding for another investigation requested by the control committee. The council decided instead to report to the local police, who closed the case without further investigation.

A major Norwegian newspaper 'Aftenposten' suddenly had on the front page a picture of whistleblower Tom Ravndal with the text that "he started to ask questions, it triggered the most scandalous construction dispute of all times on the holiday island Tjøme." This was on Friday, July 5, 2019. At least 250 construction cases at Tjøme were relevant for illegalities after Ravndal questioned the case management practice in the municipality. The police at that time had initiated investigation, and some summerhouse owners received orders from the municipality to demolish illegal parts of their properties (Lundgaard, 2019).

At the turn of the year 2017/2018, the construction scandal had already cost the municipality close to 4 million kroner (about 400,000

US dollars). The local law firm Tenden had received half of the money for their assistance along the way. One of Tenden's invoices was NOK 30,000 to process a request for access to archives in the municipality. Public broadcaster NRK gained the right to access after complaining to the county governor. At the turn of the year, the merged municipality of Færder allocated three full person-years to clean up the construction scandal. Wang (2017) wrote about the money spending in the municipality:

> The management of the municipality of Tjøme has made a huge entry into the municipal treasury to pay for services from the law firm Tenden & Co and the PR agency Zynk, allegedly because the media and outside world demanded more information than the municipality was able to handle. The residents have not benefitted much from the money spending.

The BDO (2017) review report had a very narrow mandate. Those who were subject to examination, the press, and all involved had demanded that the mandate should extend to the management of the municipal administration. However, the client for the investigation, councilor Christine Norum, would not allow it to happen. Without skepticism, the BDO investigators must have conducted their investigations in line with the client's wishes.

The new municipality Færder had two lists for cases of illegality. The public list contained 70 cases in April 2018, while it had 75 cases in September. The second list, which was not public, had 50 cases in April, while 150 cases in September 2018. The public list had cases where the municipality had contacted the owners (Blix, 2018).

Police Investigation Characteristics

Police investigation is a goal-driven and rule-restricted activity that the prosecuting authority in Norwegian police is formally heading. Police officers carry out the practical investigation methods, while the prosecuting branch in the police is in charge of everything involved with decisions about coercive measures and legal proceedings, such as the taking of evidence and evidence statements and legal claims abroad. The aim of the investigation is to elucidate whether potential offenders have transgressed various penal provisions. The investigation has to align with rules described in the Human Rights Act, the Criminal Procedure Act, and given regulations and guidelines from the Attorney General and the Directorate of the Police which belongs to the Department of Justice in Norway. The goal of the investigation is to obtain sufficiently relevant and reliable information that can clarify whether an offense has been committed, if anyone can be punished for the offense, and what punishment is appropriate in relation to the act and the person. In addition, an investigation may aim to prevent criminal circumstances from occurring and possibly, to help restore the legal status, among other things, by returning lost values (Bjerknes & Fahsing, 2018).

A professional investigation by the police should be objective. A professional investigation against a suspect will cover both what speaks in favor of the person and what speaks in disfavor of the person in question. Investigators should act as impartial as humanly possible. Knowledge and awareness of typical human error sources such as confirmation traps, tunnel vision, and social stereotypes are important. For example, gender stereotypes distinguishing women and men along traditional thinking can cause investigative errors (Brands & Mehra, 2019).

Furthermore, there is a need for a knowledge-based and transparent examination methodology. An investigation shall examine whether something has happened or will happen, and in that case what, when, where, how, why, who has participated, and what role they may have played (Bjerknes & Fahsing, 2018).

The basis for an investigation is often that the police establish suspicion that someone is planning, attempting, or has carried out a criminal offense. The police can thus in some cases start investigating before the offense is committed. The police can only start an investigation when there is a reasonable cause. The police cannot start an investigation solely based on rumors of a possible criminal offense.

The basis for suspicion in white-collar crime cases such as environmental offenses is often tips from interested parties directly to the police or via media reports. The Tjøme case was mentioned the very first time in

Hytteavisen (The Cabin News) in 2015, where Lars M. Haugen pointed out that a municipal official as building case handler might have self-interest in a case of residence requirement for a property on the Hvasser Island, a neighboring island to Tjøme Island (Brodin, 2015):

> A property on Hvasser in the municipality of Tjøme has been up for case consideration a number of times in recent years because the owners want to have the residence requirement canceled. If that happens, they can probably get a significantly higher price when they sell it. Following the municipal election, a number of politicians, in addition to, among others, Hytteavisen, received emails that more than suggest that the case-handling official must have committed illegals when he a few years ago agreed that it was not feasible to convert the property from full-year to holiday home. A majority of municipal politicians followed his recommendation.

Lars M. Haugen claimed that the case handling official, senior engineer Harald Svendsen, might have had an interest in buying the property at a favorable price (Brodin, 2015):

> Berit Aarnes has a clear memory from a meeting with you on the property in connection with carpentry work you performed. In that situation, you stated to Berit Aarnes; "I should like to live here". Therefore, you had a personal interest in taking over the property.

Berit Aarnes was one of the heirs to the property. Harald Svendsen is one of the key players in the story about corruption and building permits in the municipality of Tjøme.

An investigation is a process. Police detectives collect, process, interpret, and evaluate information in the investigation of each criminal case. The process that leads to making good investigative decisions is a pervasive search and pursuit of alternative answers without favoring the answer that police detectives already have in mind. The process most often starts with a thorough suspicion that something wrong has happened. The suspicion leads investigators to reflect on what may have happened, how it may have happened, what may not have happened, who did something, when someone did something or nothing, and why it may have happened. Various explanations should emerge that can be formulated as hypotheses in the investigation, where hypotheses are eventually confirmed or rejected. In the process, investigators must determine what information is relevant and is important to the case, and what information is not. The investigation must determine the verifiability of the information and the credibility of the sources (Solheim & Tveit, 2020).

The purpose of the investigation is to provide the necessary quantity and quality of information for the determination of the issue of prosecution, to serve as preparation for the following proceedings in court, and to avert and halt criminal offenses. There is a requirement of objectivity. If a particular person develops into a suspicious target, the investigation shall seek to clarify both the evidence against and the evidence in favor of the suspect. The presumption of innocence means that everyone is innocent until proven guilty. The prosecuting authority must convince the judges that their description of the misconduct is the only correct description. The defendant does not have to prove his or her innocence (Solheim & Tveit, 2020).

The main source of information in most police inquiries is investigative interviewing. Informants, whistleblowers, observers, suspects, and stakeholders can be relevant for interviewing to gain insights into the matter being investigated. When interviewing, Norwegian detectives determine both the reliability of the source and the quality of information. The reliability scale applied by police investigators in Norway has six levels:

1. Completely reliable source. There is no doubt about the honesty, solidity, accountability, reputation, and qualifications of the information sources.
2. Usually reliable source. While not always providing complete information, the source typically presents pieces of information that are reliable.
3. Sometimes reliable source. Most of the information from this source is trustworthy, although the misleading context or the deviant interpretations make it less reliable.

4. Usually not reliable source. There is a lack of consistency in information provided by this source that makes it hard to separate facts from fiction.
5. Completely unreliable source. The source seems to be lying most of the time, seemingly purpose.
6. Reliability cannot be assessed. The source is completely unknown, so skepticism is required toward honesty, qualifications, role, and motive.

This scale is interesting, especially the bottom level stating that an unknown and unfamiliar source is the worst origin for information in a criminal investigation. In the Norwegian culture, trust is important. When there is no explicit reason for distrust, Norwegians trust each other. In the police, however, it is different. When there is no knowledge about the source, including no reason for distrust, then police detectives should nevertheless be skeptical of information from that source.

The quality of information has a similar scale to the reliability of sources with six levels:

1. Information confirmed by several sources. Other witnesses have said the same, and documents confirm witness statements, which are typical validity checks that support the impression of acceptable quality of information.
2. Probably correct information. The information is obtained from a primary rather than a secondary source, but it seems logical and is generally consistent with other information.
3. Possibly correct information. The information is not confirmed by other independent sources but seems logical and is generally not in conflict with other information.
4. Questionable information. The information remains unconfirmed and seems in conflict with other information.
5. Unlikely information. It is not logical, as it does not follow any pattern, and it conflicts with other information.
6. The correctness cannot be judged. There is no ground to contextualize or interpret the information.

As with the reliability scale, the worst and thus least useful information is where the correctness cannot be judged. The six-by-six matrix resulting from combination of reliability and quality does not only apply to interviewees. It applies to all sources of information that are relevant to the investigation.

A difficulty for interviewers is how to overcome a suspect's resistance to cooperation and disclosure (Goodman-Delahunty & Martschuk, 2020: 152):

> In these interrogation contexts, investigators need to know what repertoire of strategies is most effective to rapidly elicit suspect cooperation and the disclosure of meaningful information, and what strategies to avoid because they are counter productive.

Open-minded interviews are more successful than accusatorial approaches in terms of interview outcomes. An interview is a source of information to solve the puzzle of crime. The goal is to obtain as much information as possible. Often, a confession does not help. Noncoercive strategies have greater effectiveness than coercive strategies as they are more likely to secure cooperation, as evidenced in a study referred to by Goodman-Delahunty and Martschuk (2020: 154):

> Strategies that diminished suspect cooperation and resulted in no response were confrontational questioning, reiteration of the caution and warnings and an emphasis on situational futility. To avoid 'no comment' or silent responses, interviewers relied on noncoercive cognitive and social strategies, such as continuing to encourage the interviewee throughout the interview, using open-ended questions, and giving suspects adequate time to respond. Suspects were more likely to cooperate when interviewers requested attention, applied rapport-building strategies, and presented evidence.

An interview can reach a desired turning point, where an initially uncooperative interviewee becomes cooperative, or the interview can reach an undesired turning point, where an initially cooperative interviewee becomes resistant or silent. Turning points are moments in time when the process changes (Goodman-Delahunty & Martschuk, 2020: 156):

> In the context of an investigative interview, a turning point attributable to the interviewer might be the moment when an initially uncooperative suspect starts to cooperate, or conversely, the moment when an interviewee who started out cooperatively shuts down and refuses to respond to further questions.

A police investigation can encompass a wide range of responsibilities. As described by Kim et al. (2020: 172), "a criminal investigation can be conceptualized as a series of information-processing tasks, such as hypothesis generation, decision-making and task prioritization." One of the police investigation's challenging aspects is that detectives often need to deal with incomplete and contradictory information to form hypotheses about what happened within a case and to prioritize actions effectively.

Whistleblowers on Tjøme Island

Lars M. Haugen's accusations in 2015 created curiosity among people who are residents on the islands of Tjøme, Hvasser, and Nøtterøy. Some summerhouse owners who had spent all their summer vacations on the islands, and who had remained loyal to laws and regulations, started to inquire about strange permits leading to new mansions on the shoreline. Ivar Amundsen and Tom Ravndal were two such summerhouse owners. They blew the whistle on Tjøme municipality in 2017, where their inquiries and detective work resulted in news coverage presented in the local newspaper. Ivar Amundsen contacted the Norwegian authority for investigation and prosecution of economic and environmental crime (Økokrim). The police agency Økokrim knew about the environmental crime case on the island of Tjøme from local police, but they had done nothing so far to investigate the case. The local police seemed reluctant to investigate the case on their own.

Ivar Amundsen got a reply from Jon Atle Georgsen and Inge Svae Grotli at Økokrim in November 2017. Amundsen and Ravndal received an invitation as witnesses to an investigative interview at the local police station on January 26, 2018. The two whistleblowers had made substantial preparations in terms of a document about 28 deviant building permit cases.

Following the four-hour-long investigative interview with Amundsen and Ravndal, the lead investigator at Økokrim, Bjørn Helge Myhre, sent the following email to both of them:

> Hello again! Thank you for a good meeting today! It was clearly informative to us, and provided a good basis for further assessments with regard to possible criminal offenses. We will now initially start by interviewing persons who can further shed light and elucidate the matter, and we aim to get started next week. For the time being, we do not want to question the most central people, as we want to have the best possible basis before it happens. We expect inquiries from the media that will be briefly commented on by stating that the South-East Police Department will get assistance from Økokrim to investigate the case, and that the police will carry out specific investigative steps without detailing them. I think that we will get back to you relatively soon if we have a need for more details in the cases that you presented to us. Now I wish you a good weekend!

The two whistleblowers had completed an impressive job by documenting 30 cases and handing over 28 of the cases on a memory stick to the police. Some of the cases had been mentioned in the media, such as case 5 about Ernst & Young executive Erik Mamelund's mansion in Nordre Svelvik beach 31, where large masses of a mountain had been blown away despite the fact that his application for approval did not mention anything about it. In the meeting with police on January 26, 2018, the whistleblowers showed pictures of the old cabin of Mamelund located on top of the mountainside at a good distance from the waterfront. The whistleblowers provided pictures of the blasting work as documentation. The informants also showed pictures of the new mansion well laid out in the terrain. The public lost the opportunity to travel along the waterfront since the mansion is now much closer to the water. In addition, surveillance cameras have been set up. The reactions to the meeting—perhaps a form of disbelief—must have been quite powerful in this matter.

The whistleblowers provided the police with useful information that Økokrim would record in their computer system analyst. Investigators can identify progress along the time axis, see relationships between people involved, and analyze transactions in context by use of the software.

Police interviewers asked Amundsen and Ravndal to pick out the six worst cases for the next meeting. The Mamelund case was an obvious candidate. It was expected that Myhre at Økokrim would contact the two informants 'relatively soon,' probably after a thorough examination of evidence so far in the police investigation. Erik Mamelund's mansion was an enormous building in the beach zone. The structure ended up in a large picture in the Norwegian daily business newspaper 'Dagens Næringsliv' on October 5, 2017, with the picture caption:

> Architect Rune Breili depicted at the summerhouse of the Norden-chief at Ernst & Young, Erik Mamelund. It increased in usable size from 146 square meters to 226 square meters after several amendment reports, but the changes never emerged as applications for new exemptions, and the changes never arrived on the politicians' table.

One strategy of deviant behavior applied by summerhouse owners with the help of the architect was to limit the initial change application, and then add incremental change applications over time. The basic rule along the Norwegian coast is that the 100 m belt is totally subject to conservation and protection against private building activities to guarantee public access and wildlife. There are, however, some exceptions. One exception is to restore the current building, but within the number of square meters already present on the private property. Another exception is to upgrade the housing standard from the previous century standard to the current century standard. However, such upgrade should preferably remain within the existing building size. In addition, external measures, such as constructing terraces, tennis courts, and swimming pools, represent serious violations of the law. The strategy thus was to send a series of incremental change notices to the municipality over time, where each incremental change represented a minor issue that the administration approved rather than treated as a case for political debate and decision.

Another strategy detected by whistleblower Ravndal was the construction of large basements that were reported as simple cavities. For example, in Ødekjær road 43 on the island of Tjøme, the owner had applied for municipal permit to build a tennis court. In the application, architect Breili had indicated that there would be a hole underneath. The detection of later architectural drawings reveals a basement with a height of 3.1 to 3.7 m with a floor area of 600 square meters underneath the tennis court. The strategy of hiding living space as underground basement had also been applied by Madsen who was sentenced to prison by the Supreme Court (2020). The owner of Ødekjær road 43 had inherited a lot of money from his late father who was a successful shipowner with the shipping company Wilhelmsen.

The media referred to Erik Mamelund as the Norden executive at Ernst & Young. It was an interesting question whether Ernst & Young as a firm thought it was acceptable to have a top manager responsible for wrongdoing and potentially illegal building activity. Therefore, an email arrived in the mailbox of the Europe executive Andy Baldwin who was above Mamelund in the firm hierarchy on March 18, 2018:

> Erik Mamelund faces suspicion of having built an illegal summerhouse on the island of Tjøme south of Oslo in Norway. The Norwegian National Authority for Investigation and Prosecution of Economic and Environmental Crime is currently investigating the matter. I wanted to bring this to your attention.

Immediately thereafter, the following reply arrived from a legal executive at Ernst & Young, Herve Labaude:

> This is to acknowledge receipt of your email to Andy Baldwin in relation with a construction matter. I am at your disposal to discuss the question further should you deem appropriate to share more specific information.

The signature in the email was Herve Labaude, deputy global general counsel—contracts/EMEIA Area counsel, Ernst & Young Global Ltd., 6 More London Place, London SE1 2DA, United Kingdom with the email address herve.labaude@uk.ey.com.

To share specific information with Labaude that he asked for, he received links to a number of media reports concerning Mamelund. Two weeks later, on April 1, 2018, the following question arrived in an email to Andy Baldwin. However, there was never a response to this email:

> Does Erik Mamelund's behavior represent a violation of your ethical guidelines?

There were several possible hypotheses about what kind of environmental crime had potentially occurred. The link to financial crime was the prospect of tremendous rise in market value of a summerhouse turned into a mansion in the beach zone. For all involved in Tjøme cases, a number of possibilities in the criminal code existed: fraud, malpractice, tax evasion, financial adultery, accounting violation, corruption, and money laundering—in addition to violations of building laws and environmental protection laws.

At the meeting on January 26, 2018, in the introduction by local police inspector Stapnes, he briefly explained that the police had several approaches. The whistleblowers expressed that they believed in fraud and tax evasion and relevant perspectives, if one failed to prove corruption. Prosecuting attorney Georgsen said at the meeting that they could not discuss such legal issues with the informants.

The whistleblowers as informants provided important evidence. The police had to find out whom this evidence might concern. Once the police had decided, charges against suspects could follow. Then, the accused will hire defense attorneys who try to prevent police evidence by means of information control as described in Chapters 4 and 5. The most resourceful in the Tjøme case—the rich and mighty on the shoreline—would probably engage the country's most skilled defense attorneys to assist them as defendants. The A team of Norwegian defense attorneys in white-collar cases is exceptionally skilled and thus terribly troublesome to the police during their investigations. Maybe that is one reason why the police had not charged anyone formally for the time being.

People in focus included summerhouse owners, municipal politicians, cabin architects, municipal bureaucrats, and construction case officials as illustrated previously in Fig. 4.2. One summerhouse owner in the figure is Stuge. Marianne Jentoft Stuge and Ulf Stuge in Nordhavna 62 received a supervisory report from the municipality of Færder on March 5, 2020. The municipality sent the review report to the owners at the address of Trosterud road 65 in Oslo. Case coordinator in the municipality, Ross Raftemo, wrote in the report that, in the municipality's opinion, violations of the planning and building act resulted from detection by inspection. Informant Amundsen made the following comment in an email March 13, 2020, about Stuge, which caused Amundsen to blow the whistle on misconduct and crime in the municipality in 2017:

> Nordhavna 62 is my neighbor in the north, Ulf Stuge. When Stuge rebuilt the old cabin in 2015, I did not object, although I noticed that this was far more than a façade change. When my neighbor two years later, however, wanted to build a large swimming pool and new annex with a new terrace 30 meters long and a terrace four meters above ground level in the direction of our cabin – then I protested. What the stupid neighbor did was to backdate the letter to his neighbors about his expansion plan by five days and set the deadline for complaints to Thursday in the middle of Easter. This was no sensible way of asking for his neighbor's goodwill toward a disturbing and for that matter questionable expansion in relation to laws and regulations. In addition, the letter to neighbors was from Breili and signed by Svendsen, who I remembered as building permits official in the municipality. Abrahamsen handled the case, who, someone told me, worked as general manager at Breili architecture firm for three years before returning to his former workplace in the municipality. This led to Tom Ravndal contacting me, and I sent a request for independence assessment of the men Svendsen and Abrahamsen in May 2017.
>
> This was the beginning of the Tjøme case. Ulf Stuge started the crisis in the municipality in collaboration with Harald Svendsen. They triggered the landslide that has raged for the last three years. Tom and I, with many good supporters, have been running an incredibly demanding and difficult process, but the collaboration has been very efficient and inspiring. We now believe that the matter is facing a rapidly escalating development. When you see what has emerged from the illegalities, and how it strikes back on Breili, Svendsen, and presumably Abrahamsen and Stegen as well – and perhaps others high up – as well as how it now turns out for Stuge even after a rather merciless review report. Then it

> might be timely to sit down and reflect whether or not own approach was wise. I might add that the review was a result of the municipality's own initiative.

Holmøy (2020) then reported about Marianne Jentoft Stuge and Ulf Stuge in Nordhavna 62:

> Applied for façade change – tore down and expanded parts of the cabin: -We have done everything we understand is right. The cabin owners in Nordhavna 62 had officials from the municipality of Færder visiting, who carried out an inspection of the leisure property, and noticed many deviations. The cabin owners in Nordhavna 62 did not build according to given permits, according to the conclusion in the municipality. In addition, three terraces, an extension, a storage room and a lower floor now existed without the necessary permission. The cabin owners do not agree with this conclusion. The mismatch between the drawings in the application, which the politicians had approved, and what eventually became the construction, became obvious during the inspection.

Whistleblowers Ravndal and Amundsen expressed early on the opinion that the police should consider charging Harald Svendsen, Petter Abrahamsen, and Hilde Stegen in the building permits department for corruption and fraud. They also felt that Rune Breili, Harald Svendsen, and Hans Petter Abrahamsen deserved charges for tax evasion due to black labor (Lundgaard, 2019). Three of the cases filed were private docks and private property issues.

According to the whistleblowers, suspicions of wrongdoing should also include some of the former and current politicians in the municipality. One of the municipal politicians was John Marthiniussen, who was the mayor from 2011 to 2015. On June 29, 2018, Færder municipality sent a letter to Marthiniussen at Mågerø road 170 on the island of Tjøme with the headline "Storage room and terrace – notice regarding enforced fine." The letter states that "pursuant to plan and buildings law §32–2, you are notified that the municipality is considering giving orders for removal of storage room, terrace and associated terrain damage" (Wien & Øystå, 2018). The following year, Marthiniussen's complaint did not succeed, because living on the seafront entails privatization of the land (Blix, 2019).

Another politician was Pål Syse, who headed the main committee for building matters in Tjøme municipality. He only reviewed meeting documents briefly and never studied them thoroughly, as he thought it would be best to have the administration make decisions based on delegation from the politicians (Holmøy, 2018):

> -Trust is the basis for the whole system, says former head of the committee, Pål Syse.

An example of a bureaucrat at that time is Christine Norum who was councilor at the head of the complete administration in the municipality. The local police treated her as a suspect when they first interviewed her. Therefore, the municipality covered her legal costs. In addition, deputy manager Gunnar Rougnø had the position of suspect when he was in his first police interview (Holmøy, 2018). As many others involved in white-collar cases do, Norum also presented herself as a victim (Schjøll, 2018):

> -It came as a shock. I felt foolish, and I felt betrayed. I had no idea, quite the contrary. It was terribly heavy. I felt depleted and quite run down, I simply got sick. I could not believe I had done anything wrong. You do not know how much you can take before everything around you fall apart. You take a risk. I have to endure this, but it hurts.

The married couple Peter Malling and Hermine Midelfart had built far too much on the Gryte Reef on the east side of Tjøme (Tilsynsrapport, 2020). People detected the construction case after an illegal explosion from cracking in connection with the start of construction in 2015. Færder municipality carried out an inspection on the property in November 2019. At that time, municipal officials discovered that the drawings submitted with the application for a temporary permit did not match the approved drawings. The deviation was so substantial that it could become relevant to demolish what the couple had built illegally. Gryte Reef is a separate island located east of Brøtsø Island. It was the local newspaper Tønsbergs Blad, which reported on the Malling and Midelfart couple in December 2019. Neither architect Rune Breili nor the couple wanted to comment on the matter (Holm, 2019). The review report from the municipality of Færder carries the date January 17, 2020, and the municipality addressed it to Hermine Elisabeth Midelfart in Hospits road 7 in Oslo. The report mentions the terrace, toilet building, and pier. Thomas Ross Raftemo, who signed the report, was the coordinator in charge of supervision and follow-up of illegality, and the report concludes as follows:

> The municipality believes that through the above descriptions, it has been shown that a number of construction activities have taken place that need permissions and dispensation from regulations. The municipality further believes that there is uncertainty about the date of building of two of the docks on the property and asks for an account of this. It is relevant for the municipality to pursue the above-mentioned circumstances in accordance with the rules of the Planning and Building Act, chapter 32. However, we allow you to make any comments and statements to the report by February 14, 2020.

Peter Malling replied to the review report from the municipality in a seven-page long letter with twenty attachments in February 2020 (Blix, 2020). Several years earlier, on July 17, 2015, the municipality of Tjøme had reported violations of the Planning and Building Act to the local police district in the city of Tønsberg:

> The municipality of Tjøme reports violations of the provisions of the Planning and Building Act, as owners have carried out illegal blasting work on the property related to the construction of a new holiday home. There was a clear condition in the building permit that the construction activities should take place without cracking. The application stated that the new building should rise on an existing foundation wall, and this assumption was obvious when the municipality granted a building permit for a new holiday home. Never did anyone apply for a change to this permit. The municipality thus requests that the guilty party(s) become subject to prosecution and punishment. We forward all documentation

> related to the construction case. Sincerely, Christine Norum, Councilor; Harald Svendsen, Senior Engineer.

The following year, the police closed the case against the contractor who blew a rock on the property belonging to businessperson Hermine Midelfart on Tjøme (Bergman, 2016):

> -The police district closed the case based on the evidence situation, because the police and prosecuting authorities have not been able to prove the necessary level of guilt, says police lawyer Øyvind Waale Hellenes in the South East police district. In July last year, the contractor Bygga Mitt Hem Inc. blew away parts of a rock on Hermine Midelfart's private islet Gryte Reef (Tilsynsrapport, 2020). The blast took place where the old cabin was, because a new summerhouse was to stand on the same foundation wall. The contractor had to put the rock masses back in place after an order from the municipality of Tjøme. Tjøme municipality believed that the owners had made an illegal intervention in the beach zone and reported the matter to the police. Now that the police have closed the case, it is up to the municipality of Tjøme if they want to impose a fee.
>
> -We have not made that decision yet. We will present the case to politicians on October 12. Then we will get an indication of what they think in this matter, says construction case manager Hans Petter Abrahamsen in the municipality of Tjøme.

Several rich and mighty on the shoreline are described above, including Erik Mamelund in Nordre Svelvik beach 31, Marianne Jentoft and Ulf Stuge in Nord harbor 62, Hermine Midelfart and Peter Malling in Gryte Reef 1, and John Marthiniussen in Mågerø road 170, all on the island of Tjøme.

The purpose of these descriptions above is to illustrate the considerable work that whistleblowers Amundsen and Ravndal had completed that helped the police when they finally initiated their criminal investigation. The whistleblowers also informed the state representative in the county. For example, they had a meeting with the state representative (county governor) on June 21, 2018, where they met Elisabeth Rui (responsible for environmental protection), Fred Ivar Syrstad (responsible for planning), and Hanna Sortland (case handling official). The purpose of the meeting was to present the systematic approach to deviant behaviors in the municipality of Tjøme. While the county governor worked case by case, the whistleblowers could paint the larger picture.

Support Team Starts Working

There is a support team at Økokrim, which has the task of assisting local police districts in cases of serious fraud, complicated financial misconduct, and environmental crime. The leader of the support team, Bjørn Helge Myhre, interviewed both whistleblowers Tom Ravndal and Ivar Amundsen. The police interviewed them as witnesses for the first time on January 26, 2018. They kept in touch as described earlier in this chapter and as reported in 'Tønsbergs Blad' (Holmøy, 2018; Olaussen, 2018).

At the end of February 2018, a person reported in writing to Økokrim. He wanted to explain the mixing of roles and about certain circumstances in Tjøme municipality. He is an auditor, and he was a deputy member of the control committee in the municipality of Færder. Økokrim invited him to testify as a witness. His case was about Brøtsø Quarry. Brøtsø Quarry is a company where Tjøme architect Breili was a shareholder together with the former Tjøme municipality and the Vestfold Festivals. The three shareholders owned 33% each. The mayor of the municipality of Tjøme was a board member, and the chairperson on the board of the Brøtsø Quarry was the general manager at Breili Architects, Svein Eriksen. The Vestfold Festivals had to close down, and the municipality of Færder wanted to withdraw from the position of shareholder. The likelihood that Breili would take over was high because Vestfold Festivals had received a lot of money, including state grants and gifts for claimed cultural projects. The money was unused in the account of Brøtsø Quarry.

The person, who reported in writing to Økokrim and then told his story in an interview with Økokrim, was former customs auditor Jarl Teig. Jarl Teig became the third important informant for the research in this book, in addition to Ravndal and Amundsen. Jarl Teig supplied on March 5, 2018, the meeting record from the control committee in the municipality of Færder where it says that the secretariat received a case on Brøtsø Quarry on January 29, 2018.

In March 2018, Tønsbergs Blad reported that the police investigation of the Tjøme case was making progress after three weeks of witness interviews (Holmøy, 2018). Head of the support team, Bjørn Helge Myhre, confirmed in the newspaper report that they were still interviewing relevant individuals. At the beginning of February, it was decided that the finance team at local Vestfold police with assistance from the support team Økokrim would investigate whether there had been corruption and/or violation of the Planning and Building Act in the Tjøme case. It was in the previous year that the three informants documented many serious errors, illegalities, and procedural errors in a number of construction cases handled by the municipality of Tjøme. The purpose of the investigation was to find out if there were grounds for criminal charges (Holmøy, 2018):

> -We will consider all circumstances. There may be several types of offenses such as breaches of trust. We need plenty of time to investigate. We need to review many documents, Myhre told Tønsbergs Blad earlier.

A few days after the witness interviews on January 28, 2018, an investigator from Økokrim called up Ravndal, and said he wanted detailed information about former construction caseworker Dag Hugo Kilvær. Kilvær worked in the municipality from March 2011 to January 2014. He replaced Abrahamsen, who at that time architect Breili employed. Ravndal was convinced that Kilvær had important information, but Kilvær did not want to disclose the information to Ravndal. Therefore, Økokrim wanted to question Kilvær.

Informant Tom Ravndal emailed on January 30, 2018, the following message to police inspector Jan Stapnes in the local police district with a copy to Bjørn Helge Myhre, the leader of the support team at Økokrim:

> To elucidate one of the issues, I suggest that you interview Steinar Helgerud. He told me about the blasting work in case five on our list, which is Nordre Svelvik beach 31. He has several pictures and can tell that the blasting work triggered several alarm calls to the municipality

> from frightened landowners nearby. Steinar Helgerud is a civil engineer and works at Fokus Buildings in Oslo. An interview with Anne Mitsem Borgersen, former politician in the municipality of Tjøme, is likely to be appropriate. She chaired a political committee for construction matters for a long period. She was a member of the committee when the matter of Mostranda Camping was up for decision-making.

Police inspector Jan Stapnes at the local police replied on February 2:

> Thank you for a good and informative presentation of the case from your point of view. In cooperation with Økokrim, we will not follow up the matter and see what comes out it. We have noticed the names listed below.

That same week it seemed that the case was about to explode in the media again. The whistleblowers had not told anyone and not mentioned Økokrim, but nevertheless public broadcasting NRK had detected that Økokrim was actively working on the Tjøme case. NRK interviewed police inspector Jan Stapnes, and NRK was to talk to his superior the following day. Stapnes confirmed in the interview that the investigation had started, and that the support team at Økokrim assisted in the work. NRK contacted some of the rich and mighty on the shoreline for comments, but they had no comments.

Leader of the support team at Økokrim, Myhre, contacted informant Ravndal on Wednesday, February 14. Myhre could then tell Ravndal that former municipal official in building matters, Dag Hugo Kilvær (2011–2014), current official in building matters, Hans Petter Abrahamsen, as well as municipal department head Margrethe Løgavlen had all participated in interviews with the police. Myhre made a new appointment with Tom Ravndal and Ivar Amundsen. Myhre asked the two informants to select a few cases that they considered most serious. They wanted to emphasize more strongly environmental crime and harm to the shoreline, in addition to possible corruption, financial misbehavior, fraud, and tax evasion when selecting the most serious cases of illegal construction work in the beach zone.

Ivar Amundsen emailed Myhre at Økokrim on February 18. He attached three recent newspaper articles from 'Tønsbergs Blad' and

'Øyene.' Amundsen wrote in his email that the articles "provide a decent insight into the pattern in Tjøme's building affairs department: Initial application is processed by the main committee. Then the matter is taken on delegation by an individual in the building permits department, with tight relations to Breili. Then there is an avalanche of change notices, which are skillfully kept away from the main committee, politicians and even the county governor, even when informing the governor is required by law in dispensation cases."

Amundsen had a telephone conversation with Myhre on February 22. Myhre could tell that the police were well on their way in their work, and that they would question several people. Among others, former building case manager Dag Hugo Kilvær was interviewed twice. Myhre said that Økokrim wanted to wait a while before talking to informants Amundsen and Ravndal again.

The support team at Økokrim attempted to establish cooperation with Færder municipality. Already on Friday, February 2, 2018, two weeks before the interview with informants, four police detectives from Økokrim visited the municipality to ask for relevant documents to clarify facts and the sequence of events in the most serious and deviant shoreline cases. Bjørn Helge Myhre with three fraud examiners from Økokrim came to Færder and asked for information about 16 specific building permits. The documents were not readily available, so municipal manager Margrethe Løgavlen and her team had to work over the weekend to send the documents to Økokrim the following Monday. Myhre told Løgavlen that the police would soon be ready to start interviewing relevant people, but they wanted first to have some basic knowledge about the most serious cases and their developments.

Løgavlen at Færder municipality ended up sending thirty cases of questionable building permits on the shoreline to Økokrim. She did not formally report the cases. She only sent them to the police at the request of the police. This difference later became an issue, as Økokrim wanted the municipality to confirm that they reported the cases because the municipality suspected white-collar crime. The municipality then replied that Økokrim could investigate independent of whether the municipality formally reported the cases. It seemed that Økokrim wanted the municipality to take a position of requesting police investigation, while the municipality wanted to stay away from allegations and accusations against municipal employees and rich vacation homeowners on the shoreline.

While not formally wanting to report cases of wrongdoing and potential crime, the municipality became very active in reviewing properties on the island at the request of Økokrim. Municipal officials visited numerous summerhouses to review what was built compared to what the applications said in municipal files. They found substantial discrepancies all over and reported their findings to Økokrim.

In addition to investigative searches by the informants and reviews by municipal officials, the media also documented their findings. After one year, the local newspaper 'Tønsbergs Blad' had published one hundred news articles about selected mansions and their owners. Journalists had spoken to owners and their attorneys, and they had interviewed neighbors and residents on the island. This was also helpful to Økokrim, although the police chose never to contact editors and journalists directly about media findings. Two journalists at 'Tønsbergs Blad,' Holmøy and Wang, won two prestigious media prizes because of their investigative journalism into the wrongdoing on Tjøme Island.

Support Team First Findings

The first important finding for the support team was a process map of how the rich and mighty had obtained all permits without following formal procedures in the municipality. The next important finding was deviations from permits—both legal and illegal permits—that had occurred. Økokrim had not yet started to focus on offenders, as the police investigators continued with the offense-based perspective.

The offense-based perspective is concerned with the criminal act in itself, drawing upon legal definitions, motives, and means (Piquero & Schoepfer, 2010). Offense-based approaches to white-collar crime emphasize the actions and nature of the illegal act as the defining agent. For example, white-collar crime is typically a non-violent offense (Berghoff & Spiekermann, 2018). Status is not included in the definition of offense-based approaches, and status is free to vary independently from the definition in most legislation, where an offense-based approach thus allows measures of status to become an external explanatory factor.

In March 2018, 'Tønsbergs Blad' reported that "The investigation of the Tjøme-complex is intensified – Økokrim has spent three weeks in interrogations" (Holmøy, 2018):

> Three weeks ago, Økokrim started interviews with people involved in the construction cases at Tjøme.
>
> -We are still interviewing relevant people, says Bjørn Helge Myhre, head of the support team at Økokrim.
>
> In the beginning of February, the police decided that the finance team of Vestfold police with assistance from Økokrim would investigate whether there has been corruption and/or violation of the Planning and Building Act in the Tjøme situation. It was last year that many serious errors, illegalities, and procedural errors were discovered in a number of construction cases handled by the former municipality of Tjøme. The purpose of the investigation is to find out if there are grounds for criminal charges.
>
> -We will consider all circumstances. There may be several types of offenses such as breaches of trust. We need plenty of time to investigate. There are many documents we need to review, Myhre told Tønsbergs Blad earlier. In the beginning of February, Økokrim investigators obtained documents from about sixteen building cases from the municipality of Færder.
>
> -Have you asked for more cases?
>
> -No, we have not done that, says Myhre.
>
> It was the municipality of Tjøme itself that last autumn asked the police to assess whether there had been criminal offenses in the building permits department, based on BDO's (2017) examination report. This summer, the audit firm conducted an investigation of the department on behalf of the municipality. In the early summer, the construction case manager at Tjøme was suspended, when NRK Vestfold brought the news that he was taking on private contracts to fill in building applications. It turned out that he had processed these applications himself. The municipality also received notifications of problematic case handling, possible incompatibility and conflicts of interest. Both the current and former building permits manager have had close ties to the architectural firm of Rune Breili, which resourceful vacation homeowners have hired for

> planning of construction work. The applications have been processed by the building permits department at Tjøme. BDO's examination revealed violations of ethical guidelines and incorrect case management practices but failed to detect corruption. In the wake of the examination, many construction cases came up that need to be reviewed again after several serious errors and illegalities were discovered, as well as procedural errors. The municipality of Færder has established its own department for supervision and follow-up of illegality. About 100 building cases handled in the old municipality of Tjøme will be reviewed. As a result of the situation, the municipality decided to employ several new construction case managers.

The two informants Amundsen and Ravndal sent a new case to the police on March 13, 2018. The case was about Lilleskag road 89. Myhre at Økokrim thanked the following day and wrote in an email to the informants:

> This seems to be among the worst cases given the scale of the building. We interviewed Abrahamsen initially yesterday and begin to get a certain picture of the practice in handling construction cases. More interviews will be conducted, and we will keep in touch as before.

Bjørn Helge Myhre, head of the support team at Økokrim, sent a new message by email to the informants on March 23:

> Thanks for new information! We are beginning to get an idea of how the general case processing has been carried out now, and we see a need for a relatively comprehensive further investigation that also includes some cabin owners. However, the investigation must be delimited, and we must therefore prioritize the most recent and most serious cases that are not outdated (especially with regard to the cabin owners). We keep in touch, but I cannot now set a time for a new meeting.

Six weeks had now passed since Økokrim started with interviews of people who had been involved in the construction cases at Tjøme. Among others, former employees of the municipality of Tjøme were questioned. Next, Økokrim wanted to talk to some of the architect's clients whose estates were troublesome. Økokrim was of the opinion that the clients as proprietors have an independent responsibility for not going beyond documented permits that they have been granted. Thus, some vacation house owners should be included in the investigation. It was not known who would be called for questioning at Økokrim. Several of the summerhouse owners that faced suspicions in connection with the Tjøme case belonged to the group of Norway's wealthiest people, partly because they were heirs, and partly because they were active front figures in finance and business.

Rich heirs are an interesting group. They have inherited lots of money and often a business as well. They may not contribute much to the business while spending much money privately (Chrisman et al., 2018; Hiller et al., 2018) on such things as a mansion on the shoreline at Tjøme. They have access to money as a resource when questioned by the municipality, but they do not necessarily have the knowledge or skill to identify and hire the best attorney for their own case.

One of the informants had telephone contact with the investigation on April 4. It was Tom Ravndal who talked to Bjørn Helge Myhre. After the telephone conversation, Ravndal sent new documentation to Myhre and Stapnes about Flekken road 62. It was Abrahamsen as general manager at the architecture firm Breili who submitted a building application in 2012. Case manager Hilde Stegen issued the starting permit in 2012 and the certificate of completion in 2014. The builder was Reidar Solheim, while the contractor was Turid Varner. Solheim built a new terrace and developed a park between the mansion and the waterfront for Varner, who was the widow after a man who had established a small shopping empire in Norway.

Økokrim had a new review of Tjøme cases with officials at the municipality of Færder on April 11. Informant Amundsen spoke to Myhre at Økokrim on April 19. Both informants Amundsen and Ravndal had expressed concern about the progress in the police investigation. Myhre assured Amundsen on the phone that Økokrim were investigating very thoroughly. Therefore, the investigation would not be complete for a while. Myhre indicated another two years before a decision regarding prosecution could be made. This was in April 2018. Two years later, in May 2020, Økokrim had decided to prosecute the architect and two building permit officials in court, and the court hearings were set for January 2021.

At the meeting with the municipality on April 11, Økokrim provided the municipality with some priorities. Mostranda Camping and Flekken road 80 had high priority, where Økokrim asked the municipality to review all relevant material. Økokrim asked the municipality to avoid issuing fines to summerhouse owners whose properties had construction work that could end up with the owners as defendants in court.

The local newspaper summarized the case of Mostranda Camping (Brevik, 2020):

- Mostranda Camping is owned by Eva and Sigmund Arnesen, and Tore and Torunn Jacobsen.
- They bought the campsite in 2015 for nine million kroner (one million US dollars) and applied to demolish and build a new one. The architectural firm Breili & Partners was commissioned to design new buildings.
- In June 2015, the politicians approved the demolition and new development of the campsite.
- Both architect and caseworker Harald Svendsen in the then Tjøme municipality informed the politicians that all measures were in line with the development plan for Mostranda Camping and did not require a dispensation.
- The development has caused reactions. Of the 15 planned rental units, only five have been built. The owners have rented three of these to themselves. Three large buildings were set up. Two of these are located at the public beach and are occupied by the owner couples.
- In November 2017, Tønsbergs Blad revealed that all the buildings at Mostranda Camping were built in violation of the development plan, contrary to what both the architectural firm and the case handling officer had claimed.
- Reviews in December 2017 revealed that the buildings have not been built and used in accordance with the drawings that were approved.
- In November 2018, Færder municipality withdrew all permits that had been granted on the wrong basis. It was emphasized that the case

is of great public interest as the campsite is located by the public and popular recreational area and the beach.

- The owners have appealed the decision without getting through. The development plan from 2003 that had been used as an argument was considered invalid because it was not approved by the municipal council.
- In December 2019, the politicians decided that the development plan from 2003 and the old regulation plan from 1968 should be repealed. The first is considered illegal and the second as deficient and outdated.
- In the spring of 2020, preparations began for a new regulation plan for Mostranda Camping.

At the spring 2018 stage in the investigation, Økokrim planned to have a combined court case with architect, municipal employees as well as summerhouse owners as defendants in court. Two years later, however, summerhouse owners faced no charges in the initial prosecution documents submitted to the district court. On behalf of several suspected summerhouse owners, clever attorneys had worked intensively on substance defense, information control, as well as symbol defense. While Økokrim was still investigating summerhouse owners when they decided to charge only an architect and two municipal employees in May 2020, they claimed officially to continue their investigations into environmental crime by the rich and mighty along the shoreline.

Returning here to the year 2018, there was a situation meeting at the local Larvik police station on Friday, May 4. Økokrim's intention was to ask the local financial crime unit to take on more responsibility for investigative steps in the Tjøme case. However, the local detectives were reluctant to take on such tasks. They claimed resource scarcity in terms of available time. However, resource scarcity in terms of available competence was more likely. Competence is a combination of knowledge, skills, attitudes, and experience. Local financial crime investigators had experience from simple economic crime. Many local investigators have an attitude that is labeled catch-of-the-day. This label indicates that local investigations are restricted to crime cases that they either can solve or close during one single day. When local detectives learned that Økokrim expected to spend two more years on the Tjøme case before reaching a decision whether to prosecute someone, then that was far outside the comfort zone of local detectives.

Bibliography

BDO. (2017). Tjøme commune—Gransking (Tjøme municipality—Investigation), auditing firm *BDO*, Oslo, Norway, 39 pages.

Berghoff, H., & Spiekermann, U. (2018). Shady business: On the history of white-collar crime. *Business History, 60*(3), 289–304.

Bergman, A. G. (2016). Saken om Midelfart-hytta henlagt (The case against Midelfart cabin closed), *NRK*. www.nrk.no, published October 1.

Bjerknes, O. T., & Fahsing, I. A. (2018). *Etterforskning – Prinsipper, metoder og praksis (Investigation—Principles, methods, and practice).* Fagbokforlaget Publishing House.

Blix, N. T. (2018). Økokrim får fortsatt mange Tjøme-saker og regner med at etterforskningen strekker seg langt ut i 2019 (Økokrim still receives many Tjøme cases and expect the investigation to last far into 2019), daily local Norwegian newspaper *Tønsbergs Blad*. www.tb.no, published November 9.

Blix, N. T. (2019). Rådmannen vil avslå eks-ordførerens søknad om bod i sjøkanten ved Mågerø: -Det innebærer en privatisering av arealene (The councilor will refuse the ex-mayor's application for a shed at the seafront on Måger island: It implies the privatization of land), local newspaper *Øyene*. www.oyene.no, published October 27.

Blix, N. T. (2020). Saken er fortsatt ikke over: Tvangsmulkten på 20.000 kroner har vokst til 200.000 for ulovlig sand (The case is still not over: The coercive punishment of 20,000 kroner has grown to 200,000 for illegal sand), local newspaper *Øyene*. www.oyene.no, published April 7.

Brands, R. A., & Mehra, A. (2019). Gender, brokerage, and performance: A construal approach. *Academy of Management Journal, 62*(1), 196–219.

Brevik, H. C. (2020). Mostranda-saken: -Uaktuelt å rive byggene (The Mostranda case: Irrelevant to demolish the buildings), daily Norwegian local newspaper *Tønsbergs Blad*. www.tb.no, published August 18.

Brodin, T. (2015). Påstår at saksbehandler har egeninteresse i bopliktsak (Claims that the caseworker has a self-interest in the property case), local newspaper for cabin owners *Hytteavisen*. http://www.hytteavisen.no/paa

staar-at-saksbehandler-har-egeninteresse-i-bopliktsak.5798489-49617.html, publisert 24. september, lastet ned 14. september 2018.

Chrisman, J. J., Chua, J. H., Breton-Miller, I., Miller, D., & Steier, L. (2018). Governance mechanisms and family firms. *Entrepreneurship Theory and Practice, 42*(2), 171–186.

Goodman-Delahunty, J., & Martschuk, N. (2020). Securing reliable information in investigative interviews: Coercive and noncoercive strategies preceding turning points. *Police Practice and Research, 21*(2), 152–171.

Hiller, D., Martinez, B., Patel, P. C., Pindado, J., & Requejo, I. (2018). Pound of flesh? Debt contract strictness and family firms? *Entrepreneurship Theory and Practice, 42*(2), 259–282.

Holm, R. C. (2019). Hermine Midelfart risikerer å måtte rive deler av hytta på Tjøme (Hermine Midelfart risks having to tear down parts of the cabin on Tjøme), Norwegian public radio *NRK Vestfold*. www.nrk.no, published September 5.

Holmøy, E. (2018). Byggesakslederen lot seg ikke stoppe av lover og regler – fjernet svaberg, lagde superanneks og grillhus med kjøkken og bad (The building case leader did not let himself be stopped by laws and rules—removed rocks, built annex and grill house with kitchen and bathroom), local Norwegian daily newspaper *Tønsbergs Blad*. www.tb.no, published December 20.

Holmøy, E. (2020). Ny anmeldelse på vei til Økokrim: -Vi har levert omfattende dokumentasjon (New report on its way to, Økokrim: We have delivered comprehensive documentation), local daily newspaper *Tønsbergs Blad*. www.tb.no, published June 25.

Kim, S., Alison, L., & Christiansen, P. (2020). The impact of individual differences on investigative hypothesis generation under time pressure. *International Journal of Police Science & Management, 22*(2), 171–182.

Klevstrand, A., & Espeland, J. M. (2020). Sa ja til 96 prosent av strandsonesøknadene (Said yes to 96 percent of the beach zone applications), Norwegian daily business newspaper *Dagens Næringsliv*, Saturday, July 4, pages 14–15.

Larssen, L. D. (2017). Tjøme kommune svidde av 200.000 kr på kommunikasjonsbyrå (Tjøme municipality spent 200,000 kroner on communication bureau), daily Norwegian local newspaper *Tønsbergs Blad*. www.tb.no, published December 19.

Lundgaard, H. (2019). Krangling i fjæra (Arguing on the shore), daily Norwegian newspaper *Aftenposten,* magazine *A-magasinet*, Friday, July 5, pages 14–25.

Olaussen, M. (2018). Hvor mye koster det egentlig samfunnet at én privilegert mann bygger ulovlig i strandsonen? (How much does it really cost society that one single privileged person builds in the beach zone?), Norwegian daily local newspaper *Tønsbergs Blad*. www.tb.no, published May 15.

Piquero, N. L., & Schoepfer A. (2010). Theories of white-collar crime and public policy. In: H. D. Barlow, & S. H. Decker (Eds.), *Criminology and public policy: Putting theory to work*. Temple University Press, chapter 10, pages 188–200.

Schjøll, A. C. (2018). Åpenhjertig eks-rådmann snakket ut i dag: Jeg følte meg lurt og forrådt Open-hearted ex-councilor speaks out today: I feel betrayed and left), Norwegian daily local newspaper *Tønsbergs Blad*. www.tb.no, published September 13.

Schjøll, A. C. (2020). Da Emira og Morten skulle fortelle om hva de jobbet med de siste årene, ble lokalet for lite (When Emira and Morten were to tell what they have been working on in recent years, the room became too small), Norwegian daily local newspaper *Tønsbergs Blad*. www.tb.no, published January 12.

Solheim, S., & Tveit, A. (2020). *Bruker norske etterforskere skriftlig etterforskningsplan og hva tenker de om denne utviklingen i praksis? (Do Norwegian investigators use a written investigation plan and what do they think of these developments in practice?)* Master thesis, experience-based master's degree in investigations, Oslo, Norway: Politihøgskolen (Norwegian Police University College).

Supreme Court. (2020). Case HR-2020-1353-A (sak nr. 19-158426STR-HRET), *Norges Høyesterett (Norwegian Supreme Court)*, Oslo, Norway, June 26, 2020.

Tilsynsrapport. (2020). Gryteskjær fritidsbebyggelse tilsynsrapport (Leisure Building Supervision Report), Færder kommune (Municipality of Færder), reference 19/49893, dated January 17, 4 pages, postmottak@faerder.kommune.no.

Wang, M. (2017). Informasjon eller tåkelegging: Det er bare to grunner til at kommunen ringer PR-rådgiver (Information or obfuscation: There are only two reasons why the municipality calls pr consultants), Norwegian daily local newspaper *Tønsbergs Blad*. www.tb.no, published December 20.

Wien, H., & Øystå, F. (2018). Folk blir hengt ut som lovbrytere (People are hung out as criminals), Norwegian magazine for municipalities *Kommunal Rapport*. www.kommunal-rapport.no, published July 6.

8

Tjøme Investigation Progress

At this time in 2018, nobody was charged. A distinction is made in the Norwegian criminal justice system between suspected, charged, indicted, convicted, and lawfully convicted person. Many individuals were in the spring 2018 suspected based on rumors, allegations, and accusations. To charge someone, the police needs a solid basis for suspicion. When they have charged someone, the police have powers available to them, such as search warrants, seizure of documents, and wiretapping. The next step of indictment after a charge means that the suspect becomes a defendant during court hearings. A conviction can be appealed, while a lawfully convicted person has received the final verdict. When Økokrim later changed the position of architect Breili from suspected to charged individual, they searched his home and his office.

Two architects who had observed the activities of architecture firm Breili for a while handed in a complaint to the local police on May 21, where they accused Breili of fraud. The alleged fraud was to write construction applications that the architecture firm should know violated the law. Nevertheless, the architecture firm did it to profit from their rich clients.

P. Gottschalk, *Detecting and Investigating Environmental Crime*,
https://doi.org/10.1007/978-3-030-74184-6_8

Please recall that former customs auditor Jarl Teig had detected Brøtsø Quarry. He emailed Myhre at Økokrim on May 30, 2018, and expressed his concern related to the former mayor's involvement with architect Breili. Myhre replied the same day:

> Hello to you too! Thank you for interesting information!

Informant Ravndal spoke again to Myhre on the phone on June 6. Myhre told him the current situation in the investigation:

- Three construction case handlers have been questioned, and two more are scheduled for questioning. Interviews with summerhouse owners have started.
- The police focus on a few big cases that are fresh and rough.
- The Tjøme case has high priority at Økokrim, and substantial resources are used.
- It is very relevant to give the municipality corporate punishment.
- Økokrim had again a meeting with local police on June 5 in Larvik, where they planned further progress.
- Økokrim had a meeting with Færder municipality on June 4. It was a positive meeting, but Økokrim would not interfere in political matters. The politicians were discussing financial sanctions against summerhouse owners, and they were discussing abandonments and withdrawals of building permits.
- Økokrim would prefer formal reports from the municipality to the police where suspicions of crime are documented. They liked the formal report by the two architects who formulated allegations of crime against architect Breili.
- Økokrim has received several inquiries from both the public and neighbors regarding construction matters.
- Økokrim also looks at building cases from other contractors and architects.
- Økokrim identified quickly that municipal employees had committed service neglect, and the investigation has not exactly weakened this perspective on the Tjøme case.
- This will all take time.

After the summer break, the two informants were to meet again with Økokrim on August 23. The meeting was to take place at the new police station in Tønsberg. Just before the summer break, Bjørn Helge Myhre contacted Tom Ravndal. Myhre asked whether there might be deviant building cases in the former Nøtterøy municipality as well that implied corruption. Ravndal could confirm that it was the case by providing a few examples.

On August 29, Tom Ravndal testified formally as a witness at 10 am. He testified by explaining formally his knowledge of five deviant building permit cases that Myhre prioritized. It was Myhre alone who arranged the formal witness interview. Afterward, Ravndal reported that the questioning went well. At that time, all formal interviews in the Norwegian police took place by means of video recording where both sound and picture were stored. To turn the witness statement into a document, Myhre wrote down text from the recording. Then, Ravndal was to sign the witness statement.

In the autumn of 2018, Økokrim seemed very interested in prioritizing the Tjøme case, while the local police district was more skeptical. It was still officially the situation that the South East Police District was responsible for the investigation with the assistance of Økokrim.

Ravndal had a new meeting with Økokrim on Friday, September 14, where he signed the explanation. The day before, Ravndal had another meeting with the county governor as the state representative in Vestfold.

Because of Økokrim's growing interest in the Tjøme case as environmental crime, combined with the local police district's reluctance and incompetence, the criminal investigation was transferred from the local police to the national police in October 2018. Police inspector Stapnes and prosecutor Øyvind Waale Hellesnes in the local police district withdrew from the case, while senior prosecutor Tarjei Istad at Økokrim took over the role of senior investigating officer. Although the environmental perspective was growing in importance, corruption was still the main topic in the police investigation.

It was more than a year later when Madsen in Grimstad had been convicted of environmental crime (Agder, 2019), as described earlier in this book, that Økokrim formally made the Tjøme case a matter of environmental crime. Økokrim was probably unsure to what extent emerging environmental crime importance would work better or worse in court compared to traditional white-collar offenses. The traditional perspective would be corruption, where architect Breili was accused of bribing municipal employees. The corruption perspective could be extended to include Breili's clients as summerhouse owners. However, the choice of accusing involved individuals of environmental crime rather than purely financial crime might seem more appropriate and more promising, since environmental protection is of growing concern in the public as well as by the legislative authorities.

Investigative Thinking Styles

Økokrim conducted more interviews in the fall of 2018, which included former councilor Christine Norum and former department head Gunnar Rougnø. While interviewing more and more people from the municipality, Økokrim continued the dialogue with Ravndal and Amundsen as whistleblowers and informants. They had a new meeting on Monday, September 24. At every meeting, Økokrim wanted to know specifics about prioritized cases that the informants had detected. Now, Myhre at Økokrim was very much on the corruption path involving architect Breili and municipal official Svendsen. Svendsen had recently involved himself in obtaining municipal approval for Erik Bakkejord from Oslo to build a new summerhouse of 400 square meters in Langvika on Tjøme (Wang, 2018).

When contacted by the media, senior prosecutor Tarjei Istad at Økokrim said that the criminal investigation would probably last beyond 2019. Nobody was charged, he said, but there were several suspects and witnesses. By leaving the architect, municipal employees, and summerhouse owners in the position of suspects rather than charged persons, the individuals had no rights related to the police investigation. When the police charge someone, then the police have access to more methods in the investigation, such as searching premises and seize evidence. At the same time, the charged person has the right to access to and insight into the police investigation. By hiring an attorney, Økokrim can deny the attorney insight when working for a suspect, but Økokrim cannot deny the attorney insight when working for a charged person. Since attorneys tend to apply substance defense, information control, and symbol defense that can seriously disturb and damage police investigations, it can be smart to leave the position of a person as a suspect for a time period that is as long as possible.

Senior prosecutor Tarjei Istad was again interviewed in November 2018 (Blix, 2018):

> Several investigators at Økokrim are now working full-time on the Tjøme case and around 25 construction cases have been looked at so far. Earlier this autumn, Økokrim took over responsibility for the investigation of the Tjøme case. Tarjei Istad in the environmental department at Økokrim is responsible for potential prosecution of cases against offenders.
>
> -We are now looking at whether errors have been made in the handling of construction cases at Tjøme, errors that are so many and so serious that a criminal liability can be considered. There must be a minimum scope and certain seriousness before we can talk about criminal liability. We have looked at a variety of cases, around 25 cases so far, and we think there may be more addressed, Istad tells the local newspaper 'Øyene'. There are still cases coming in to Økokrim. The municipality of Færder expedites further cases that they perceive as very serious in their view. Private individuals also file cases, Istad has learned.

It is interesting to reflect on investigative thinking styles at this stage of the investigation. Dean and Gottschalk (2007) make a distinction between how investigators prefer to think about solving crime. They argue that the most difficult thing about an investigation is learning to think how to think about an investigation. The key, therefore, to be a successful investigator, no matter what type of personality or kinds of abilities, lies in what style of investigative thinking that investigators bring to the case. A style is a preferred way of thinking. A style is not an ability. That is, an ability refers to how someone can do something, while a style refers to how someone likes to do something.

Dean and Gottschalk (2007) make a distinction between the method thinking pattern, the challenge thinking pattern, the skill thinking pattern, and the risk thinking pattern. The nature of method thinking is systematically going forward in predetermined steps. The steps are collecting, checking, considering, connecting, and constructing. Detectives follow the protocol according to formal rules.

There are few signs of the method thinking pattern in the Tjøme investigation, since the police was reluctant to take the Tjøme case on and make it their own. Instead, both local police and Økokrim initially were most of the time passive receivers of whatever information they could get. They adapted their thinking style to whatever thinking style was evident among the whistleblowers as informants, municipal officials as sources for archival material, and media outlets as storytellers. While the method style is predetermined, structured, and rigorous, police investigators in the Tjøme case had initially no plan or no structure. They simply kept an open mind toward whatever people would tell them. In that way, they avoided the famous tunnel view in police investigations, where one goes into a trap of believing what is first told or detected. Rather than completing formal investigative steps simply for the purpose of completing those steps, police officers listen to what they were told, so that they later could frame their learning in terms of hypotheses.

The challenge thinking pattern is more appropriate as a label for the Tjøme investigation. While passive at first, police investigators did find it challenging and motivating to listen to all people that were worried about the rich and mighty occupying the shoreline by means of money and networks. The challenge style is experiential and experimental rather than formal and procedural. The challenge might be victim driven as everyone has noticed that public access to attractive recreational areas deteriorates quickly. The challenge might be criminal driven where detectives might find it unfair that the rich and mighty can do whatever they like. Selmini (2020: 6) describes the challenge thinking pattern in the following way:

> The challenge thinker dedicates all his energies to the investigation. The rest of the world, and particularly his private life, disappears; what matters is doing justice. For some it is an obsession, for others a passion.

In meeting this investigative challenge, detectives require skill to relate and communicate effectively with a variety of people to obtain information to establish a workable investigative focus. Such skill also requires detectives to be flexible in the way they approach people and the case, while maintaining an appropriate level of emotional involvement toward victims, witnesses, informants, and suspects. Certainly, on-the-job experience has a large role to play in shaping the investigative thinking of a challenge. As detectives conduct interviews and archival searches, they become driven by the intensity of the challenge, which motivates them to do the best job they can for the victims by catching the criminals and solving the crime. This challenge style of thinking is all about what motivates detectives. Initially, Økokrim was passively supporting the local police. Then, they got more motivated to conduct interviews on their own, and the next thing that happened was that Økokrim got so motivated that they took on the case by themselves and stepped up the resources involved in the investigation.

With the skill style, detectives are most concerned with how they relate to people. They know that people are the most important source of information. If people do not want to tell, or do not want to tell the truth, then police investigators are lost in pieces of text in documents and messages that they do not understand. Any piece of information must be understood in a context that people can explain to detectives. What people remember is dependent on how interviewers ask questions. Twenty years ago, Norwegian police changed their tactics completely in interviews. They moved from interrogation to interviewing. They moved away from confrontation to cooperation, where any witness or suspect is treated as a valuable source of information. They moved away from confession to avoidance of confession, where the feeling of guilt will only manipulate information provided in an interview. Detectives at Økokrim seem to fit this aspect of the skill thinking pattern as they successfully made informants feel welcome and important in the investigation.

Dean and Gottschalk (2007) argue that conducting simple investigations often only requires some combined application of the three styles of investigative thinking presented so far:

- Following the basic steps and logical stages in the application of the method style of thinking.
- Dealing appropriately with the challenges that arise when investigating, by identifying the likely intensity sources in this style of thinking.

- Using one's understanding of the importance of the skill style of thinking to effectively relate to people involved with the particular investigation and get the necessary information to build the case is, generally, all that is usually required of a detective in simple, non-complicated investigations to get a reasonable result.

However, in more serious and complex investigations, this is where the need for the fourth investigative thinking style of risk comes into prominence. This risk style of investigative thinking has at its core a preference for engaging in proactive efforts of justifiable risk-taking behavior to increase the chances of getting a good result in an investigation. There are three justificatory lines that are concerned: the lawfulness of the proposed risk, how logical the risk is, and the degree of latitude afforded to the risk both economically and conceptually. Each of these three criteria needs to be justifiable for the risk to be worth it. There is, of course, no point in proceeding with an investigative risk if it is not covered by the law in terms of the powers granted to the police. The three key processes involved in the risk style of investigative thinking are discovery–creativity–development. Creativity is located in between the processes of discovery and development in this investigative thinking style to indicate its pivotal nature in both other processes.

In 2018, police investigators at Økokrim had not entered the risk style in the Tjøme case. They stayed safe in the skill and challenge styles of thinking. Maybe, they would need to move into the risk style if they were to investigate the rich and mighty with their top-level attorneys. When the attorneys intensify their substance defense, information control, and symbolic defense, police detectives must respond in creative and risky manners. At the same time, they must watch that the attorneys cannot catch the police in making mistakes neither procedural nor factual. As will become evident in the following description of the years 2019 and 2020, Økokrim avoided emphasis on the rich and mighty and thus stayed away from the risk style of investigative thinking.

Avoiding Mansion Owners

Returning to the fall of 2018, when Tarjei Istad was interviewed in November, he emphasized that the investigation would take time. He argued that each construction incident lived its own life, where crime might be evolving as they were speaking. He said that they were currently two police detectives in addition to himself as a police lawyer, who worked permanently on solving the Tjøme case. They needed for sure next year, 2019, to get to the bottom of this case (Blix, 2018):

> -I do understand that people want clear answers now, but we have the case and need to conceptualize our work. We must learn more. Although we talked to the South East police district on an ongoing basis when they had the case, we also need to read all available material. I think we need to look at all issues with neutral and objective eyes and evaluate them no matter how others have done it, that is my attitude toward this. Istad believes the transfer of the case to Økokrim results in more resources and time.
>
> -In general, our privilege is that we choose what cases to take on, and we can work completely dedicated to those cases. Everyday life in police districts is quite different, they must take on whatever is urgent at the same time, Istad explains. The politicians in the technical main committee also had questions about the progress of the investigation when they met with us earlier this week. Municipal manager Margrethe Løgavlen had announced that they had a meeting with Økokrim last week.
>
> -We then handed over new incidents to Økokrim. They start over again, but have much more resources, and there is a team with previous experience and expertise in the field. But it will take time; that is what it will do, Løgavlen told the politicians.

While Istad answered politely about the local police district in the newspaper interview, he was very upset and dissatisfied with work done in the police district. He told the two informants that Økokrim had to take on the investigation to make sure that it was professional. They had operationalized the investigation, such as reviews of money transfers and bank accounts belonging to involved individuals and institutions. Istad also complained that former Tjøme municipality had made so many mistakes that suspected summerhouse owners might easily escape prosecution. He said in November 2018 that Økokrim's ambition was to proceed to court hearings in 2021, which they did.

We now move into the year 2019, where Istad at Økokrim had completed reviews of several construction matters. The question in March was whether there had been such serious errors in the case processing that Økokrim should consider criminal liability. The construction cases reviewed came from the BDO (2017) report, from the municipality, and from private persons. Økokrim put all of them in a pile and looked at them with fresh eyes, regardless of who they came from.

The latter seems a little astonishing, since Norwegian police is trained in considering both the reliability of the source and the quality of the information. The reliability scale applied by police investigators in Norway has six levels as presented previously: completely reliable source, usually reliable source, sometimes reliable source, usually not reliable source, completely unreliable source, and reliability cannot be assessed. Please recall that this scale is interesting, especially the bottom level stating that an unknown and unfamiliar source is the worst origin for information in a criminal investigation. In the Norwegian culture, trust is important. When there is no explicit reason for distrust, Norwegians trust each other. In the police, however, it is different. When there is no knowledge about the source, including no reason for distrust, then police detectives should nevertheless be skeptical of information from that source. When Istad at Økokrim said in March 2019 that they looked at all construction cases with fresh eyes, regardless of who they came from, then that might be a polite answer to a journalist, but not necessarily the case. For example, construction cases from the municipality were likely to receive more attention than cases from the informants or the BDO report that turned out to be wrong.

The quality of information has a similar scale to the reliability of sources with six levels: information confirmed by several sources, probably correct information, possibly correct information, questionable information, unlikely information, and the correctness cannot be judged. As with the reliability scale, the worst and thus least useful information is where the correctness cannot be judged. When the former customs auditor reported about potential misconduct in the ownership of Brøtsø Quarry, then both the source reliability and the information quality could be assigned to level six. It was outside the scope of the investigation, and the person provided more accusations and allegations and less facts and substance.

The milestone in the criminal investigation in March 2019 was to consider whether detected misconduct and wrongdoing were serious enough to press criminal charges. If Økokrim would conclude that the evidenced seriousness was not enough, then Økokrim could close the Tjøme case. They did not. The errors were beyond what might be excused or explained in some plausible manner. The errors were mainly done by the architect Breili and the two municipal employees Abrahamsen and Svendsen.

When Istad was asked about the mansion owners who had initiated construction work without valid permits from the municipality, he replied again that it was up to the municipality to file reports to the police. He wanted to avoid mansion owners as long at the municipality did not formulate allegations and accusations against them. The municipality commented that they did not need to accuse mansion owners, as Økokrim could on their own initiate an investigation against them.

Istad could in June 2019 tell the press that Økokrim currently had four suspects. Three worked in the municipality, the fourth person did not. He did not know that the police had ever gone through so many building cases to review municipal work processes. He said he had hoped to complete the investigation before the summer, but that would not happen. He said Økokrim needed everything well documented if they were going to file a criminal case for public service failure.

Tønsbergs Blad asked again Istad about the police investigating the mansion owners in the beach zone. The prosecutor replied that the police will not investigate those people if no formal reports were received from the municipality of Færder. The councilor in the municipality of Færder had then informed the main political body that cases were not reported because Økokrim previously had stated that the police did not need formal complaints to investigate.

Thus, the police were waiting for the municipality to act, while the municipality was waiting for the police to act at the end of June 2019. Both wanted to avoid taking the initiative against the rich and mighty on the shoreline and their attorneys. The municipality had intensified their review work by visiting several properties and submitted review reports to the owners. Owners were informed that they had to remove terraces, ports, extensions, and other illegal constructions. They were told to open again for the public along the shoreline. No more permits were granted, and permits given illegitimately were withdrawn. The municipality argued that this was relevant work for them. Criminal perspectives were not relevant for them. That was work for the police, the municipality argued.

It was a surprisingly passive attitude on both sides, even though the detected wrongdoing was quite significant violations of the Planning and Building Act that could be punished with incarceration. At the same time, the suspicion of corruption had become stronger. When the municipality found reason to overturn more and more building permits with a high risk of being sued by rich heirs and others in the beach zone, it was strange that the municipality did not file formal reports to the police.

One hypothesis could be that Økokrim would first test evidence in cases against the four suspects before the police possibly seized summerhouse builders. There was a rumor that Svendsen spoke like a waterfall during police interviews, and that could match information about an increase in case complexity. The questioning of a person who had bought property from Breili or Svendsen apparently provided new insights. The new information and the recent interviews might have contributed to the fact that Økokrim would wait with any charges till the fall of 2019.

Another hypothesis could be that mansion owner Madsen in Grimstad was not yet finally convicted. The final conviction at Agder court of appeals came on September 16, 2019. The question of guilt was then finally determined. However, Madsen had appealed the sentence of prison for 45 days. While the district court had decided that she did not have to go to prison, the court of appeal determined that she had to go to prison. The Norwegian Supreme Court was in 2020 to make a final decision regarding the sentence. Since most of the mansion cases on Tjøme Island were like the Madsen case, Økokrim was probably waiting for the final court outcome in Grimstad before charging mansion owners on the island of Tjøme.

Økokrim continued in the summer of 2019 to investigate four suspects. The four suspects were architect Breili, former building permit officials Svendsen and Abrahamsen, and former municipal manager Gunnar Rougnø.

The final verdict for Hanne Madsen from the Supreme Court (2020) in Norway arrived on June 26, 2020. The court upheld the sentence from previous courts of 45 days in prison. While the court of appeals had divided the days into 21 days in prison and 24 days depending on behavior, the Supreme Court decided that she had to spend all the forty-five days in prison (Berglihn, 2020).

Search of Suspect Property

In the fall of 2019, a new person became visible in the Tjøme investigation. It was Tore Høviskeland, who oversaw the environmental crime unit at Økokrim. He had successfully prosecuted the environmental crime case against mansion owner Madsen in Grimstad (Agder, 2019). Please recall that Hanne Madsen had taken over a business enterprise from her father. She bought a property behind her summerhouse, which was within the one-hundred-meter belt as well. When acquiring the land, she told the municipality of Grimstad that she would do nothing to it, so that the public could continue walking across into a public beach nearby. However, few years later, she had expanded the summerhouse and built an annex. She had built an underground hallway from the summerhouse to the annex. On the acquired property, she built a tennis court (DN, 2019).

Høviskeland seemed ready to take on the Tjøme case, as he encouraged the municipality to report wrongdoing to the police. But the municipality was still reluctant to file formal reports to the police. The reason now was that the municipality wanted to define themselves as a victim of wrongdoing by mansion owners. They did not want to move into the position of an offender who potentially accused mansion owners without enough evidence. They did not want the municipality to be sued by the rich and mighty and their attorneys.

The position of a victim is attractive. As argued throughout this book, both convicted offenders, such as mansion owner Madsen, and suspected offenders, such architect Breili, want to be perceived as victims. The position of victim is not only attractive for individuals. It is also attractive for organizations. As argued by Gottschalk and Benson (2020), scandals develop from exposure to investigation, where organizational accounts often change over time. The municipality of Færder wants to deny wrongdoing in the former municipality of Tjøme. They prefer scapegoating where some employees are isolated for blaming and attribution of guilt. Attribution of guilt is a frequent issue when something goes wrong. Attribution theory (Keaveney, 2008) and the blame game hypothesis (Lee & Robinson, 2000) suggest that establishing causality for a negative event is concerned with efforts by suspects to assign misleading attribution to others (Eberly et al., 2011). Self-blame is rare and often non-existent. Nobody will blame oneself for a negative event. Self-blame is attributing a negative event to one's behavior or disposition. Pontell et al. (2014) found that some are too powerful to blame.

In this perspective of attribution and blame for wrongdoing in the former municipality of Tjøme, the rich and powerful argue that the municipality made mistakes and is thus to blame. The municipality in turn argues that there were some rotten apples, such as Svendsen and Abrahamsen, in the municipality. They deny that the municipality as such as a rotten apple basket, rotten apple barrel, rotten orchards, or rotten apple field. The rotten apple approach suggests that when the municipality removes the few rotten apples from the organization, then everything might be fine again. A rotten-barrel approach is more serious for the organization, as there is a deviant business culture that involves top executives and that is difficult to change (Punch, 2003).

The top executives in the municipality of Tjøme during the wrongdoing were Christine Norum as councilor and Bente Bjerke as mayor. They were both happy with the BDO (2017) examination conclusion that misconduct but no crime had occurred. The leader of the control committee in the municipality at that time was Dag Lars Erichsrud. He was not happy with the quality of the BDO examination. He wanted an extended investigation by KPMG, but the main political body denied funding of it. Instead, Norum and Bjerke spent NOK 200,000 on a public relations campaign to tell the story of an innocent Tjøme municipality.

Depending on the outcome of the offender versus victim battle, the mansion owners or the municipality end up in the position of victim. If mansion owners are victims, then they can claim compensation from the municipality for the delays and problems involved in completing their construction work. If the municipality is the victim, then they can claim reversal of construction work and punish mansion owners with substantial financial replacements and fines.

Høviskeland asked for a meeting with the municipality that took place on August 28. Økokrim was represented by Høviskeland, Istad, Myhre, and five police detectives. Councilor Eeg, municipal manager Løgavlen, and illegality controllers Raftemo and Løvold represented Færder municipality. The following day, councilor Eeg informed the politicians about the meeting. She said that the municipality would now file criminal cases formally and officially to the police. Obviously, Høviskeland had successfully convinced the councilor that formal reports were required and justified. He had argued that if the municipality does the basic work of criminal profiling, then Økokrim has a much better chance of collecting evidence that will be enough to get convictions in court.

In October 2020, Økokrim searched at several addresses in the Tønsberg area in Vestfold. A search is different from a razzia. While a razzia implies that the police do not know what they are looking for, a search is directed at certain predefined objects and observations. The police needed a search warrant from the court, where the court requires that someone is formally charged before they issue a search warrant. Økokrim had charged four people. None of them were summerhouse owners. The four persons were entitled to defense lawyers paid by the state, and the defense lawyers were entitled to review police documents.

While Økokrim investigators were busy interviewing and searching, the municipality of Færder was busy preparing more Tjøme cases to file as police reports. Municipal officials first visited a suspected property. Then, they reviewed relevant documents. Then, they sent a notice to the owner, and let the owner respond. Then, they issued a statement regarding illegal constructions and missing permits. Finally, they had a construction case that could formally be filed as a police report.

Bibliography

Agder. (2019). *Court case 18-178766AST-ALAG*, September 16, 2019, Agder lagmannsrett (Agder court of appeals), Agder, Norway, 22 pages.

BDO. (2017). Tjøme commune – Gransking (Tjøme municipality—Investigation), auditing firm *BDO*, Oslo, Norway, 39 pages.

Berglihn, H. (2020). Hanne Madsen må i fengsel: -Viktig at det reageres strengt (Hanne Madsen must go to jail: It is important that the reaction is strict), daily Norwegian business newspaper *Dagens Næringsliv*. www.dn.no, published June 29.

Blix, N. T. (2018). Økokrim får fortsatt mange Tjøme-saker og regner med at etterforskningen strekker seg langt ut i 2019 (Økokrim still receives many Tjøme cases and expect the investigation to last far into 2019), daily local Norwegian newspaper *Tønsbergs Blad*. www.tb.no, published November 9.

Dean, G., & Gottschalk, P. (2007). *Knowledge management in policing and law enforcement*. Oxford University Press.

DN. (2019). Ny dom kan skremme pengesterke hytteeiere fra ulovlig bygging – Ny dom gir god hjelp til kommuner som vil stoppe ulovlige byggearbeider i strandsonen (New court sentence might scare money-rich cottage owners from illegal construction building—New court sentence provides good help to municipalities who want to stop illegal construction work in the beach zone), daily Norwegian business newspaper *Dagens Næringsliv*. Saturday, September 21, page 2.

Eberly, M. B., Holley, E. C., Johnson, M. D., & Mitchell, T. R. (2011). Beyond internal and external: A dyadic theory of relational attributions. *Academy of Management Review, 36*(4), 731–753.

Gottschalk, P., & Benson, M. L. (2020). The evolution of corporate accounts of scandals from exposure to investigation. *British Journal of Criminology, 60*, 949–969.

Keaveney, S. M. (2008). The blame game: An attribution theory approach to marketer-engineer conflict in high-technology companies. *Industrial Marketing Management, 37*, 653–663.

Lee, F., & Robinson, R. J. (2000). An attributional analysis of social accounts: Implications of playing the blame game. *Journal of Applied Social Psychology, 30*(9), 1853–1879.

Pontell, H. N., Black, W. K., & Geis, G. (2014). Too big to fail, too powerful to jail? On the absence of criminal prosecutions after the 2008 financial meltdown. *Crime, Law and Social Change, 61*(1), 1–13.

Punch, M. (2003). Rotten orchards: "Pestilence", police misconduct and system failure. *Policing & Society: An International Journal of Research and Policy, 13*(2), 171–196.

Selmini, R. (2020). Exploring cultural criminology: The police world in fiction. *European Journal of Criminology*, 1–17, published online https://doi.org/10.1177/1477370820939362.

Supreme Court. (2020). Case HR-2020-1353-A (sak nr. 19-158426STR-HRET), *Norges Høyesterett (Norwegian Supreme Court)*, Oslo, Norway, June 26, 2020.

Wang, M. (2018). Teknologiriking vil bygge fritidsbolig på 400 kvadratmeter (Technology rich wants to build vacation home of 400 square meters), Norwegian local newspaper *Tønsbergs Blad*. www.tb.no, published September 23.

9
Tjøme Investigation Conclusion

Construction scandals had caused the business crisis in the municipality of Tjøme that the municipality of Færder inherited. Scandals are publicized instances of transgression that run counter to social norms. Scandals typically result in condemnation and discredit and other consequences such as bad press, disengagement of key constituencies, the severance of network ties, and decrease in performance (Piazza & Jourdan, 2018). A scandal can be an act of elite deviance that might include financial, physical, and morally harmful behavior committed by privileged members in society. A crisis from scandals can be a fundamental threat to the organization, characterized by ambiguity of cause, effect, and means of resolution (Bundy & Pfarrer, 2015; König et al., 2020).

When the year 2020 started, the police investigation had been going on for two years. Four persons were charged, but not indicted. They were all individuals who had helped the rich and mighty to get what they wanted on the shoreline. The rich and mighty were not charged. However, the municipality of Færder had taken on the role of filing reports to the police regarding the rich and mighty. So far, the municipality had completed and submitted four cases to the police that were

P. Gottschalk, *Detecting and Investigating Environmental Crime*,
https://doi.org/10.1007/978-3-030-74184-6_9

concerned with illegal construction work. The illegal aspect was either that there was no permit given at all, or that permits were obtained in an illegitimate manner (Holmøy, 2020).

Whistleblower Ravndal had now become a permanent and influential informant for Økokrim. Ravndal was working almost full time on detecting illegal construction activities on the island of Tjøme. Ravndal had retired from his job a few years ago. Now, he had time to search municipal files, visit construction sites, talk to neighbors, and discuss findings with investigative journalists. He had meetings with Økokrim in August and November in 2019, and again on March 11 in 2020. He was told at the meeting that Økokrim had decided to indict architect Rune Breili and former building department employee Harald Svendsen for corruption. Furthermore, they would indict former building department employee Hans Petter Abrahamsen for serious negligence in public service.

Formal Prosecution Decision

In addition to Høviskeland, the leader of the environmental crime department, and Istad from the same department, another senior Økokrim prosecutor attended the meeting in March. The person was Esben Kyhring who was assigned the task of prosecuting the corruption case in court against defendants Breili and Svendsen. Another attorney at Økokrim would prosecute the service negligence case against Abrahamsen.

After planning the prosecution of these three defendants in court, Økokrim started to investigate the accusations against mansion owners more seriously. They let the owners remain with the position of suspects to avoid involvement of suspects' attorneys. Four cases of mansion owner wrongdoing had been reported by the municipality of Færder so far, in addition to more than a dozen from the whistleblowers.

Økokrim indicted formally three persons on Thursday, April 30, 2020, as the police agency stated on their website okokrim.no that "Architect and building manager in former Tjøme municipality indicted for gross corruption":

> Økokrim has brought charges against two persons for serious corruption. In addition, a third person has been required to pay a fine for service errors in connection with the processing of building cases in the beach zone.

The third person was Abrahamsen who decided not to accept the fine and thus instead defend himself in court in January 2021 together with Breili and Svendsen. The fine was NOK 15,000 for grossly breaching the duty of service. Abrahamsen decided on building applications for the beach zone in violation of the Planning and Building Act according to the indictment.

The statement on the website okokrim.no continued:

> A former building manager in the then Tjøme municipality, as well as an architect, has been indicted for gross corruption. According to the indictment, the architect must have performed services at a total value of approximately NOK 50,000, which has benefitted the municipal building manager. By virtue of this position in the municipality, the building manager was responsible, among other things, for processing building applications from the architectural firm for which the architect worked.
>
> This year's threat assessment from Økokrim points out, among other things, that Norwegian municipalities are vulnerable to corruption, and that the danger of corruption is particularly present where the private sector meets the public sector.
>
> -Corruption in the public sector is particularly important to combat. Environmental crime is often driven by financial profit, says first state attorney Hans Tore Høviskeland, head of the environmental crime division in Økokrim.
>
> -The case complex has been extensive and applies to illegal building decisions in the beach zone. Illegal construction in the coastal zone leads to irreversible interference with nature and restricts public access to coastal traffic. It also affects the natural diversity, says Høviskeland. The South East Police District started investigating the cases in 2017. Økokrim took over the investigation in October 2018.

The media contacted several stakeholders to get their comments. Former mayor Bente Bjerke had no comment, while current mayor Jon Sannes Andersen emphasized that nobody was convicted so far. He expressed satisfaction that a new step was entered after two years of police investigation.

In May, councilor Toril Eeg forwarded to Økokrim two additional formal cases regarding mansion owners. Økokrim confirmed that they now were focusing on summerhouse owners after completing the indictment of the architect and two former employees. Høviskeland said that two summerhouse owners were of interest at this stage. Both owners were rich heirs.

Investigating Mansion Owners

One of the two summerhouse owners that Økokrim investigated was Fredrik Selvaag, who had inherited a contracting company for apartment buildings. He was mentioned earlier in this book as he wrote a personal letter to a political committee in the municipality with the help of his attorney Mari Helen Gulsvik at law firm Selmer. His property at Nautnes road 129 on Tjøme Island was demolished and expanded without valid permits from the municipality. Selvaag demolished the old cabin that was both small and old fashion without facilities that most people expect to have today. Selvaag as the heir spent some of his inherited money on architect Breili's plan to build a mansion on the same property. However, no permits or dispensation had been granted. Not only was the new mansion several hundred square meters, Selvaag had also built a one-hundred-and-fifty square meter terrace.

The only construction work normally allowed after removal of an old cabin is to build a new cabin of the same size on the same spot. Even that requires formal application and formal approval from the municipality. Selvaag with the help of Breili had completely ignored the municipality. Therefore, the municipality denied Selvaag and his family to spend time in the new mansion in the summer of 2020. Nobody can move into a new building without municipal approval of completeness of the building in Norway.

Architect Breili argued that they had not at all ignored the municipality. In fact, he had invited politicians to visit the property. The visiting politicians were informed about the new summerhouse, he claimed.

However, there are no documents in municipal files regarding demolition and construction in Nautnes road 129. The heir Selvaag has argued that he left the whole project to architect Breili, and that he had no way of controlling whether Breili had submitted applications according to municipal requirements.

Selvaag applied the principal-agent perspective where the principal is unable to control what the agent is doing. In this perspective, Selvaag is the principal, while Breili is the agent. A principal is often unable to control an agent who does work for the principal. The principal-agent model assumes narrow self-interest among both principals and agents. The interests of principal and agent tend to diverge, they may have different risk willingness or risk aversion, there is knowledge asymmetry between the two parties, and the principal has imperfect information about the agent's contribution (Bosse & Phillips, 2016; Chrisman et al., 2007; Pillay & Kluvers, 2014; Williams, 2008).

The other summerhouse owner that Økokrim investigated was Celestine Clauson. She is also mentioned earlier in this book. Anne Tellefsen, her attorney, claimed that the amount of 20 million Norwegian kroner represented what Clauson had paid for expansion rights when she bought the summerhouse (Holmøy, 2020). Clauson paid 30 million Norwegian kroner for the summerhouse, but it was only worth 10 million without the expansion rights. Clauson's attorney claimed that the seller had told Clauson that the municipality had granted expansion rights. The claimed expansion approvals included a swimming pool, a tennis court, a larger basement, as well as changes to the façade of the building. When the whistleblower informed politicians in the municipality that approvals were not according to rules and regulations, the councilor recommended to the politicians to determine permits and approvals as illegitimate and not valid. The reaction from Tellefsen was to send a long letter to the politicians telling that the municipality would be liable for Clauson's loss of NOK 20 million kroner. One of the politicians expressed in the media that she perceived the letter as a threat (Holmøy, 2020). The threat was symbolic, because the municipality said it is never involved in the private transaction of real estate between a seller and a buyer (Holmøy, 2020).

When Celestine Clauson in 2015 bought the summerhouse in Nesholm road 67 on the island of Tjøme from the family Kittelsen, who had owned the place for generations, she was only 18 years old. Her father, Conrad Clauson, was the real owner. He had made a fortune as a broker when he worked for Credit Suisse First Boston in London, where he exploited bank clients for his personal benefit (Ahuja, 2005):

> Credit Suisse First Boston has started legal proceedings against a former member of staff after he failed to repay a $750,000 loan. CSFB has launched the claim, which was filed in the UK high court in February, against Conrad Clauson, a Norwegian who worked as a private client broker at Donaldson, Lufkin & Jenrette, which became part of Credit Suisse in August 2000.

Clauson had placed his fortune in a trust in the tax haven Virgin Island (Berglihn et al., 2007). In the Tjøme case, he first paid architect Breili successfully to apply for permits in the municipality. Then, he paid attorney Tellefsen to fight for the illegal permits by symbolic defense (Holmøy, 2020). In 2020, father and daughter were investigated by Økokrim.

My Lecture in the Municipality

As explained in the introduction to this book, the whistleblowers and informants did not only keep journalists in the media and investigators in the police updated on their findings. They also kept me as a researcher updated by a continuous stream of emails over several years. From time to time, the local newspaper 'Tønsbergs Blad' and the local station of Norwegian public broadcasting interviewed me about the Tjøme case.

In an email on February 3, 2020, I was invited to give a lecture on the Tjøme case in the municipality of Færder. I accepted the invitation and gave my lecture on March 5. I emailed the title of my presentation in advance, as well as forty-two presentation pictures. "Rich and mighty in the beach zone – Convenience for white-collar offenders by environmental crime" was the title of my talk. I started by presenting the theory of convenience with its three dimensions of motive, opportunity, and willingness. Then, I moved into the Tjøme case and said it would indeed be problematic to collect enough evidence to get some of the rich and mighty convicted. For example, they could blame the municipality, and their skilled attorneys could successfully apply various defense strategies.

I predicted that the architect and the building officials would change their position from being charged to being indicted by Økokrim. This did indeed happen two months later. I also predicted that Økokrim might be reluctant to charge and indict anyone of the rich and mighty who without valid permits had done all sorts of construction work on their properties. The sixty-five persons present in the municipal hall at Færder reacted to my prediction. Several persons in the audience found it terrible if I was right. There was no equality for the law, they said, as the rich and mighty were the main beneficiaries of law violations. People thanked me afterward for my analysis.

The following day, a journalist from the local newspaper 'Tønsbergs Blad' called me and said that a politician in the audience was very skeptical of my presentation. The journalist interviewed me, and then, the newspaper published the following story (Lien & Brevik, 2020):

> Professor Petter Gottschalk was the main attraction when politicians in Færder invited him to an open meeting on environmental crime and cases of illegality on Thursday evening. The meeting was held in the municipal council hall at Tinghaug in Færder. Organizers write on Facebook that there were 65 audience members in the hall. The attendees were private individuals, local politicians, and employees from the administration in the municipality.
>
> Anders Mathisen (right-wing party) was one in the audience. He reacts strongly to what he heard from the speaker.
>
> -Allegations and accusations were made about named persons, employees in the municipality, summerhouse owners, firms and attorneys. That was totally unacceptable, says Mathisen. According to Mathisen, Gottschalk was harsh in criticizing the municipality.
>
> -Some asked if he could refer to similar cases from other municipalities in Norway. Then he pointed to Fredrikstad municipality and said that the municipal director had resigned after pressure, which Gottschalk thought was the only right thing to do. He probably suggested that the same should be done in Færder, says Mathisen. Mathisen has also reactions directed at representatives from the ruling parties who laughed and

supported Gottschalk's naming of employees and private persons in the municipality (..).

Professor Gottschalk finds the criticism too far-fetched, especially because Mathisen left the meeting before it ended.

-He claims that I have called former and current municipal employees corrupt. I never said that. My interest was triggered when 'Tønsbergs Blad' contacted me and asked for a comment on the Tjøme case. Then I said that it might be corruption. I never said anyone was corrupt. The three persons that I mentioned by name are also named by Tønsbergs Blad in this case, but corrupt I never called them.

According to Gottschalk, his main message at the meeting was that there is no equality for the law, because his empirical studies show that caseworkers and architects can be prosecuted and convicted, while wealthy and influential summerhouse owners hire very expensive lawyers who get the out of trouble (...)

Tone Kalheim and Jørn Magdahl, two left-wing politicians, react to the claims by Mathisen.

-Petter Gottschalk was invited to give a lecture based on his work and his authorship, with a general and overall approach. He is of course aware of what is happening in our municipality and follows this with interest, but should not emphasize individual matters, which he also did not, says Magdahl and Kalheim to Tønsbergs Blad. The two tell that they have great respect for the job the municipality is doing to follow up the cases of illegality (...)

There were people in the audience with great breadth and affiliation. The comments and questions asked from many different angels were factual and good and were answered by the lecturer. When a capacity like Petter Gottschalk manages to lecture about the complex Tjøme case with enthusiasm and dedication, we as listeners quickly become equally engaged of course. An excellent lecturer can manage to gain the attention of the audience, and sometimes recognition of reality triggers spontaneous reactions. It cannot and should not be controlled.

-There were 65 listeners present, which confirms the topic's timeliness and interest, they say.

The right-wing politician Mathisen raised the issue of my lecture also in the next meeting in the municipality. Jon Sannes Andersen, the mayor of Færder, who listened to my talk, commented at the meeting that everything I said did indeed make a lot of sense. While the right-wings had ruled during the scandalous construction activities until 2019, the left-wings won the election in September 2019. Mayor Andersen was a member of the labor party.

Preparing for Court Hearings

The district court in Horten took on the Tjøme case. Horten is a city in the Vestfold and Telemark County north of the city of Tønsberg and the Færder islands. The court had reserved six weeks starting January 19, 2021. This book is about the stage of police investigation and will thus not include the court hearings or the outcome from the court. Whether the defendants are acquitted or convicted in the first instance of a district court is not important in Norway. It is obvious that so much prestige is at stake for all involved in the Tjøme case that the criminal case will end up in a court of appeals the following year 2022. If the case is appealed again and admitted to the Supreme Court, then a final verdict cannot be expected until 2013.

A court hearing scheduled for six weeks is a long time. One of the time-consuming activities will be presentations of exclusive mansion properties on the island of Tjøme. Økokrim was hoping for at least thirteen properties where the municipality had completed their reviews and submitted the cases formally to Økokrim. The construction cases should illustrate how the administration approved new summerhouses, basements, docs, and parks without required political concessions. In June 2020, Økokrim had so far received four property cases.

While Økokrim prepared for the court hearing directed at three defendants, an architect and two former employees in the municipality, Økokrim obviously wanted to spend time in court to prepare for a next round of court hearings with summerhouse owners as defendants at a later point in time.

In addition to the heirs Selvaag and Clauson, Økokrim also wanted to investigate properties belonging to Erik Mamelund and Hermine Midelfart. If they found enough evidence, and if the first case in Horten district court would end in convictions in 2021, then Økokrim would proceed by indictment toward four summerhouse owners with likely court hearings in 2022. Please recall that Ernst & Young executive Mamelund's mansion in Nordre Svelvik beach 31 had caused large masses of a mountain to be blown away. This happened even though his application for approval did not mention anything about changing the landscape. In the meeting with the police on January 26, 2018, the whistleblowers showed pictures of the old cabin owned by Mamelund located on top of the mountainside at a good distance from the waterfront. The whistleblowers provided pictures of the blasting work as documentation. They also showed pictures of the new mansion well laid out in the terrain. The public lost the opportunity to travel along the waterfront since the mansion is now much closer to the water. In addition, surveillance cameras have been set up. Mamelund's mansion was a dominating building in the beach zone. As reported and pictured in the Norwegian daily business newspaper 'Dagens Næringsliv' on October 5, 2017, the summerhouse increased its usable living area from 146 square meters to 226 square meters after several amendment reports.

Hermine Midelfart was also presented earlier in this book. The married couple Peter Malling and Hermine Midelfart had built far too much on the Gryte Reef on the east side of Tjøme (Tilsynsrapport, 2020). Neighbors detected the construction case after an illegal explosion from cracking in connection with the start of construction in 2015. Færder municipality carried out an inspection on the property in November 2019. At that time, municipal officials discovered that the drawings submitted with the application for a temporary permit did not match the approved drawings. The deviation was so substantial that it could become relevant to demolish what the couple had built illegally.

While police detectives and fraud examiners at Økokrim started on the Selvaag, Clauson, Mamelund, and Midelfart cases, prosecutors at Økokrim prepared for court hearings half a year later. Prosecutor Kyhring focused on the corruption case involving Breili and Svendsen, while another prosecutor at Økokrim focused on public service negligence involving Abrahamsen. Høviskeland at Økokrim focused on environmental crime that might be prosecuted against Selvaag, Clauson, Mamelund, and Midelfart in 2022. Økokrim attorney Høviskeland said in an email to whistleblower Ravndal on June 29, 2020, when Ravndal had congratulated Økokrim with the conviction of white-collar offender Madsen:

> Thank you!!! Yes, this was an important verdict and well that we were fully upheld in the claim. It was preventively important. The verdict will also have an impact on the further work on the cases from Tjøme!

The whistleblowers Amundsen and Ravndal continued their efforts to detect and communicate findings. For example, on June 19, 2020, they had a meeting with councilor Eeg and municipal manager Løgavlen regarding facts about more suspects in addition to the four summerhouse owners already reported to the police (Holmøy, 2020). The whistleblowers were important sources of information. As suggested by Lynch and Stretesky (2013: 6), "citizens can act as important adjuncts in the enforcement of laws" through "community participation in environmental policing."

Investigative journalists in the media continued their search for new scandals, comments, and reviews. Most prominent was the local newspaper 'Tønsbergs Blad' with its journalists Aunemo (2018, 2020), Brevik (2020a, 2020b), Holmøy (2017, 2018, 2019, 2020), Larssen (2017), Lien and Aunemo (2019), Lien and Brevik (2020), Olaussen (2018), Schjøll (2018), Skavhellen and Lien (2018), and Wang (2017, 2018, 2019). While this local newspaper reported piecemeal on cases focusing on construction work lacking permits, the Norwegian business newspaper 'Dagens Næringsliv' reported more on the offenders and their positions in society (Dugstad & Skaalmo, 2017; Heltne & Klevstrand, 2019; Holth & Winge, 2018; Klevstrand, 2018; Klevstrand & Randen, 2018; Klevstrand & Solem, 2020; Klevstrand et al., 2020; Ytreberg, 2018). Norwegian public radio brought some reports (Bergman, 2016; Holm, 2019, 2020), and a magazine for municipalities mentioned the scandal a few times (Wien & Øystå, 2018). The major Norwegian newspaper 'Aftenposten' also mentioned the Tjøme scandal a few times (Lundgaard, 2019). A special newspaper published only once a week for the local islands 'Øyene' was careful in reporting potential law violations (Blix, 2018, 2019, 2020). The cited articles are just examples of the stream of articles in the mentioned media. It was likely that the stream would continue for the rest of 2020 and increase in volume as the court hearings would start in January 2021. However, since this book is concerned with the stage of police work in terms of criminal investigation, it is beyond the scope of this book to follow the Tjøme scandal any further.

The final verdict from Norwegian Supreme Court (2020) on June 26, 2020, in the case against Hanne Madsen on the shoreline in Grimstad municipality must have been encouraging for preparing for court hearings in the Tjøme case. Madsen had to go to jail, and so did the person responsible for applying and constructing on her behalf (Berglihn, 2020). Both principal (Hanne Madsen) and agent (Tor Øystein Michaelsen) had to go to jail. The Tjøme scandals seem very similar indeed, where Breili was the agent in most of them. Principals included Selvaag, Clauson, Mamelund, and Midelfart. Please recall that lack of oversight and guardianship might result from agency problems suggesting that a principal is often unable to control an agent who does work for the principal. The principal-agent model assumes narrow self-interest among both principals and agents. The interests of principals and agents tend to diverge, they may have different risk willingness or risk aversion, there is knowledge asymmetry between the two parties, and the principal has imperfect information about the agent's contribution (Bosse & Phillips, 2016; Chrisman et al., 2007; Pillay & Kluvers, 2014; Williams, 2008). In the Grimstad case with Madsen and Michaelsen, however, the Norwegian Supreme Court (2020) considered both guilty and incarcerated both for 45 days.

While encouraging for Økokrim, they could also notice differences between the successfully prosecuted Grimstad case and the forthcoming Tjøme case that disfavor the latter. The main difference was the role and behavior of the municipalities in their case handling. Grimstad municipality made no mistakes. Tjøme municipality made several mistakes. For example, caseworkers in the municipality had approved construction applications where political bodies had the actual decision rights. While architect Breili had helped the rich and mighty use the incremental method of applications, caseworkers reviewed each marginal application in isolation. Prosecuting individuals such as Selvaag, Clauson, Mamelund, and Midelfart could thus be more risky than prosecuting

Madsen, since Tjøme defendants might argue that they had indications and elements of approvals and permits, and that they are not responsible for mistakes in the municipality.

In July 2020, Færder municipality was reviewing a fifth construction incident to report to Økokrim. The shipping heir Wilhelmsen had inherited money and bought the property Hønsø road 73 for 20 million kroner (US$ 2 million) in 2014. He wanted to remove the building on the property and build a luxury summerhouse of 311 square meters. He did, but the building was never approved, as the application for permit did not match the summerhouse that municipality officials found when they visited the property.

Bibliography

Ahuja, V. (2005). CSFB sues former broker over $750,000 loan. *Financial News London*. www.fnlondon.com, published August 8.

Aunemo, E. W. (2018). Her får Tønsbergs Blad-journalistene pris for Tjøme-sakene (Here the Tønsberg Blad journalists receive a prize for the Tjøme cases), daily local Norwegian newspaper *Tønsbergs Blad*. www.tb.no, published Wednesday, September 12.

Aunemo, E. W. (2020). Advokaten til millionærarving Celestine (22) krever at skattebetalerne skal betale honoraret hennes (The millionaire heiress attorney demands that the taxpayers pay her fees), daily local Norwegian newspaper *Tønsbergs Blad*. www.tb.no, published January 19.

Berglihn, H. (2019). Princess-arving dømt til fengsel (Princess heir sentenced to prison), daily Norwegian business newspaper. *Dagens Næringsliv*, Friday, September 20, page 4.

Berglihn, H. (2020). Hanne Madsen må i fengsel: Viktig at det reageres strengt (Hanne Madsen must go to jail: It is important that the reaction is strict), daily Norwegian business newspaper *Dagens Næringsliv*. www.dn.no, published June 29.

Berglihn, H., Gjernes, K., & Nordahl, B. O. (2007). *Metoderapport: Skjult i skatteparadis* (Method report: Hidden in tax haven), daily Norwegian business newspaper *Dagens Næringsliv*. www.dn.no, published January 18.

Bergman, A. G. (2016). Saken om Midelfart-hytta henlagt (The case against Midelfart cabin closed). *NRK*. www.nrk.no, published October 1.

Blix, N. T. (2018). Økokrim får fortsatt mange Tjøme-saker og regner med at etterforskningen strekker seg langt ut i 2019 (Økokrim still receives many Tjøme cases and expect the investigation to last far into 2019), daily local Norwegian newspaper *Tønsbergs Blad*. www.tb.no, published November 9.

Blix, N. T. (2019). Rådmannen vil avslå eks-ordførerens søknad om bod i sjøkanten ved Mågerø: -Det innebærer en privatisering av arealene (The councilor will refuse the ex-mayor's application for a shed at the seafront on Måger island: It implies the privatization of land), local newspaper *Øyene*. www.oyene.no, published October 27.

Blix, N. T. (2020). Saken er fortsatt ikke over: Tvangsmulkten på 20.000 kroner har vokst til 200.000 for ulovlig sand (The case is still not over: The coercive punishment of 20,000 kroner has grown to 200,000 for illegal sand), local newspaper *Øyene*. www.oyene.no, published April 7.

Bosse, D. A., & Phillips, R. A. (2016). Agency theory and bounded self-interest. *Academy of Management Review, 41*(2), 276–297.

Brevik, H. C. (2020a). Tom er ikke overrasket (Tom is not surprised), daily Norwegian local newspaper *Tønsbergs Blad*. www.tb.no, published April 30.

Brevik, H. C. (2020b). Mostranda-saken: Uaktuelt å rive byggene (The Mostranda case: Irrelevant to demolish the buildings), daily Norwegian local newspaper *Tønsbergs Blad*. www.tb.no, published August 18.

Bundy, J., & Pfarrer, M. D. (2015). A burden of responsibility: The role of social approval at the onset of a crisis. *Academy of Management Review, 40*(3), 345–369.

Chrisman, J. J., Chua, J. H., Breton-Miller, I., Miller, D., & Steier, L. (2018). Governance mechanisms and family firms. *Entrepreneurship Theory and Practice, 42*(2), 171–186.

Dugstad, L., & Skaalmo, G. (2017). Slik fikk han de rikes hytter godkjent på Tjøme (This is how he got the cabins of the rich approved on Tjøme), daily Norwegian business newspaper *Dagens Næringsliv*. www.dn.no, published September 29.

Heltne, L., & Klevstrand, A. (2019). Her står Frederik Selvaags hytte ferdig – og ubrukelig. Fredrik Selvaag har fått nei fra Færder kommune til å ta i bruk sin nye Tjøme-hytte. Kommunen ber i stedet om en ny strandsone-søknad for hele hytta (Here stands Fredrik Selvaag's cabin finished—and useless. Fredrik Selvaag got no from Færder municipality to move into his new Tjøme cabin. The municipality asks instead for a new shoreline application for the complete cabin), daily Norwegian local newspaper *Tønsbergs Blad*. www.tb.no, published July 17.

Holm, R. C. (2019). Hermine Midelfart risikerer å måtte rive deler av hytta på Tjøme (Hermine Midelfart risks having to tear down parts of the cabin on Tjøme), Norwegian public radio *NRK Vestfold*. www.nrk.no, published September 5.

Holm, R. C. (2020). To personer tiltalt for grov korrupsjon i hyttekommune (Two persons charged with serious corruption in cabin municipality), Norwegian public radio *NRK Vestfold*. www.nrk.no, published April 30.

Holmøy, E. (2017). Lederen av kontrollutvalget på Tjøme: Betenkelig det som har forekommet (Leader of the control committee on Tjøme: Worrysome what has taken place), local Norwegian daily newspaper *Tønsbergs Blad*. Wednesday, October 4, page 6.

Holmøy, E. (2018). Byggesakslederen lot seg ikke stoppe av lover og regler – fjernet svaberg, lagde superanneks og grillhus med kjøkken og bad (The building case leader did not let himself be stopped by laws and rules—removed rocks, built annex and grill house with kitchen and bathroom), local Norwegian daily newspaper *Tønsbergs Blad*. www.tb.no, published December 20.

Holmøy, E. (2019). Denne terrassen skal bli mindre – må rive ulovlig del (This terrace shall become smaller—must remove illegal part), daily Norwegian local newspaper *Tønsbergs Blad*. www.tb.no, published December 27.

Holmøy, E. (2020). Ny anmeldelse på vei til Økokrim: Vi har levert omfattende dokumentasjon (New report on its way to, Økokrim: We have delivered comprehensive documentation), local daily newspaper *Tønsbergs Blad*. www.tb.no, published June 25.

Holth, F., & Winge, N. K. (2018). Bekymringer etter Tjøme-saken (Worries after the Tjøme case), Norwegian daily business newspaper *Dagens Næringsliv*. Tuesday, May 22, page 20.

Klevstrand, A. (2018). Nedgraderer Tjøme-arkitekt (Downgrade Tjøme architect), Norwegian daily business newspaper *Dagens Næringsliv*. Wednesday, July 20, page 17.

Klevstrand, A., & Espeland, J. M. (2020). Sa ja til 96 prosent av strandsonesøknadene (Said yes to 96 percent of the beach zone applications), Norwegian daily business newspaper *Dagens Næringsliv*. Saturday, July 4, pages 14–15.

Klevstrand, A., & Randen, M. (2018). Suzuki importør bygget Tjøme-luksus på overtid: Risikerer å måtte rive hytta (Suzuki importer built Tjøme luxury on overtime: Risks having to tear down the cabin), Norwegian daily business newspaper *Dagens Næringsiv*. Wednesday, July 25, page 1.

Klevstrand, A., & Solem, L. K. (2020). Økokrim undersøker hytteeiere, tiltaler arkitekt (Økokrim investigates cabin owners, charge architect), Norwegian daily business newspaper *Dagens Næringsliv*. Saturday, May 2, pages 12–13.

Klevstrand, A., Solem, L. K., Dugstad, L., & Skaalmo, G. (2020). Tjøme-arkitekt Rune Breili tiltalt for grov korrupsjon – ga arkitekt-råd for 50 000 kr (Tjøme architect Rune Breili indicted for serious corruption—gave architect advice for 50,000 kroner), Norwegian daily business newspaper *Dagens Næringsliv*. www.dn.no, published April 30.

König, A., Graf-Vlachy, L., Bundy, J., & Little, L. M. (2020). A blessing and a curse: How CEOs' trait empathy affects their management of organizational crisis. *Academy of Management Review, 45*(1), 130–153.

Larssen, L. D. (2017). Tjøme kommune svidde av 200.000 kr på kommunikasjonsbyrå (Tjøme municipality spent 200,000 kroner on communication bureau), daily Norwegian local newspaper *Tønsbergs Blad*. www.tb.no, published December 19.

Lien, A. O., & Aunemo, E. W. (2019). Hun får betalt av fleire som er dypt involvert i Tjøme-skandalen. Nå har hun samlet byggebransjen mot kommunen (She gets paid by many who are deeply involved in the Tjøme scandal. Now she has gathered the construction business against the municipality), daily Norwegian local newspaper *Tønsbergs Blad*. www.tb.no, published May 24.

Lien, A. O., & Brevik, H. C. (2020). Frp-Anders forlot møte i protest: -Det ble rett og slett for drøyt (Right-wing Anders left the meeting in protest: It simply became too much), daily Norwegian local newspaper *Tønsbergs Blad*. www.tb.no, published March 9.

Lundgaard, H. (2019). Krangling i fjæra (Arguing on the shore), daily Norwegian newspaper *Aftenposten,* magazine *A-magasinet*, Friday, July 5, pages 14–25.

Lynch, M. J. (2020). Green criminology and environmental crime: Criminology that matters in the age of global ecological collapse. *Journal of White Collar and Corporate Crime, 1*(1), 50–61.

Lynch, M. J., & Stretesky, P. B. (2013). The distribution of water-monitoring organizations across states: Implications for community environmental policing and social justice. *Policing: An International Journal of Police Strategies & Management, 36* (1), 6–26.

Olaussen, M. (2018). Hvor mye koster det egentlig samfunnet at én privilegert mann bygger ulovlig i strandsonen? (How much does it really cost society that one single privileged person builds in the beach zone?), Norwegian daily local newspaper *Tønsbergs Blad*. www.tb.no, published May 15.

Piazza, A., & Jourdan, J. (2018). When the dust settles: The consequences of scandals for organizational competition. *Academy of Management Journal, 61*(1), 165–190.

Pillay, S., & Kluvers, R. (2014). An institutional theory perspective on corruption: The case of a developing democracy. *Financial Accountability & Management, 30*(1), 95–119.

Schjøll, A. C. (2018). Åpenhjertig eks-rådmann snakket ut i dag: Jeg følte meg lurt og forrådt Open-hearted ex-councilor speaks out today: I feel betrayed and left), Norwegian daily local newspaper *Tønsbergs Blad*. www.tb.no, published September 13.

Skavhellen, Ø., & Lien, A. O. (2018). Tønsbergs Blad vant pris for dekningen av byggeskandalen på Tjøme (Tønsbergs Blad won prize for reporting the construction scandal), Norwegian local daily newspaper *Tønsbergs Blad*. www.tb.no, published May 2.

Supreme Court. (2020). Case HR-2020-1353-A (sak nr. 19-158426STR-HRET), *Norges Høyesterett (Norwegian Supreme Court)*, Oslo, Norway, June 26, 2020.

Tilsynsrapport. (2020). Gryteskjær fritidsbebyggelse tilsynsrapport (Leisure Building Supervision Report), Færder kommune (Municipality of Færder), reference 19/49893, dated January 17, 4 pages, postmottak@faerder.kommune.no.

Wang, M. (2017). Informasjon eller tåkelegging: Det er bare to grunner til at kommunen ringer PR-rådgiver (Information or obfuscation: There are only two reasons why the municipality calls pr consultants), Norwegian daily local newspaper *Tønsbergs Blad*. www.tb.no, published December 20.

Wang, M. (2018). Teknologiriking vil bygge fritidsbolig på 400 kvadratmeter (Technology rich wants to build vacation home of 400 square meters), Norwegian local newspaper *Tønsbergs Blad*. www.tb.no, published September 23.

Wang, M. (2019). Mer trøbbel for hyttemillionær. Kommunen mener alle gamle tillatelser er ugyldige (More trouble for cabin millionaire: The municipality says that all old permits are invalid), Norwegian local newspaper *Tønsbergs Blad*. www.tb.no, published December 5.

Wien, H., & Øystå, F. (2018). Folk blir hengt ut som lovbrytere (People are hung out as criminals), Norwegian magazine for municipalities *Kommunal Rapport*. www.kommunal-rapport.no, published July 6.

Williams, J. W. (2008). The lessons of Enron: Media accounts, corporate crimes, and financial markets. *Theoretical Criminology, 12*(4), 471–499.

Ytreberg, R. (2018). Kommunen stoppet Breili-prosjekt på Tjøme (The municipality stops Breili project on Tjøme), Norwegian daily business newspaper *Dagens Næringsliv*, Monday, April 30, pages 4–5.

10
Environmental Crime Perspectives

This chapter presents some critical criminology perspectives on environmental crime. While traditional street crime is characterized by offense, offender, and victim that are reflected in laws, environmental crime as white-collar offenses lacks many of the obvious crime characteristics. For example, environmental wrongdoing can or cannot be punishable. Identification of victims is dependent on harms that are not acceptable. Research needs to address how environmental crime is conceptualized and how criminalization processes develop. Certain mechanisms are interesting to study, which cause some of the harms from environmental wrongdoing to be legal while others are criminalized. Green criminology has become an important and expanding field in criminology, particularly within the perspective of critical criminology, which focuses on challenging traditional understandings and uncovering false beliefs about criminal justice.

P. Gottschalk, *Detecting and Investigating Environmental Crime*,
https://doi.org/10.1007/978-3-030-74184-6_10

Punishing Environmental Crime

The case studies from Norway presented in this book suggest that it is indeed possible to get offenders convicted of environmental crime if police investigations follow an iterative path. Police investigators are dependent on informants that can lead to evidence relevant for punishing environmental crime. The case of Madsen in Norway shows that the move from a lower court to a court of appeal and then the Supreme Court did not change the verdict (Agder, 2019; Supreme Court, 2020). She had to go to prison for a tennis court and other illegal construction work on her property in the beach zone.

Billiet et al. (2014) studied several environmental crime cases in Belgium as the cases moved from lower courts to the court of appeal. They analyzed judicial policy lines concerning the punishment of environmental crime using a Belgian dataset of individual crime cases. Public prosecutors in Belgium decide to prosecute in about eight percent of the environmental crime cases that reach their office. They prioritized offenses that have, or might have, serious consequences for public health and the environment, have an organized crime character, are committed in a professional context, or concern the exploitation of a facility or activity without the required environmental permits.

This priority is interesting for comparison to Norway. While the Madsen case was a single offender with a single offense (Agder, 2019; Berglihn, 2020; DN, 2019), the Tjøme case has elements of organized crime in a professional context (Holmøy, 2020a, 2020b; Wang, 2019; Ytreberg, 2018). The professional context was the architectural firm that was financed by some of the mansion owners on the island. The architectural firm was run by architect Breili, who was the link between wealthy summerhouse owners and municipal officials who approved construction work. The indications of organized crime were the numerous incidents of illegitimate permits over time that had the same pattern. For example, a series of applications were used to document incremental changes and to avoid the complete construction project to become a matter of municipal approval or disapproval in political committees.

Another interesting result from the Belgian study is the distribution of sanctions for environmental crime (Billiet et al., 2014). The total number of convictions for environmental crime was 1,023. The dominating sanction was a fine, which occurred in 985 instances (87%). Prison sentence only occurred in one single case (0.1%). A combination of prison sentence and fine was passed to offenders in 102 instances (10%). The remaining offenders were convicted to community service (3%).

While no comparable statistic is available from Norway, it is nevertheless interesting to note that fine was the main sanction historically. Almost exclusively in the past, the rich and mighty on the shoreline received a fine. However, one instance made the public prosecution review the sanctioning regime. It was a rich entrepreneur who had an illegal swimming pool on the shoreline. The local municipality fined him. He accepted the fine and has been paying the fine ever since. The fine was 5,000 kroner per month (US$500) that the offender easily could pay, since it represents an expense of only 60,000 kroner (US$6,000) annually, which is a minor amount for the offender. When this story became publicly known, and politicians and the criminal justice system realized that fines do not work against wealthy offenders in Norway, then prison sentence, eventually combined with a fine, became the right path forward for criminal sanctions. Madsen was sentenced to 45 days in prison and a fine of 100,000 kroner (US$10,000).

The Belgian convictions mentioned above refer to courts of first instance. While 1,023 defendants received their final verdict in the first instance, 142 defendants received their final verdict in a court of appeal. The distribution of criminal sanctions is similar for the two court levels: 87 and 82% were fined, 0.1 and 2% got prison sentence, 10 and 12% were incarcerated, and 3 and 6% suffered other sanctions (Billiet et al., 2014).

In their research, Billiet et al. (2014) formulated several hypotheses that they tested empirically. For the courts of first instance, they found a highly significant negative impact of the fraction of proven accusations compared to the total number of accusations on the probability of being acquitted. To prosecute a defendant successfully, the fraction of proven accusations needs to be beyond a certain threshold level. Individuals had a lower probability of acquittal when offending during professional activities and when prosecuted for an offense against environmental permit conditions.

Furthermore, Billiet et al. (2014: 482) identified several determinants for the severity of criminal sanctions:

> The effective sanctions inflicted on individuals were more stringent when sanctioning prioritized offenses, in cases causing negative health impacts or involving civil parties, for breaches of the duty to have an environmental permit, waste offenses and offenses causing odor hindrance, and when the offender had a record of previous convictions or acted from financial motives.

Again, a comparison to the Norwegian case studies is relevant, as we do not know the outcome of the Tjøme case in court. Certainly, there were breaches of the duty to have an environmental permit in the sense of permission to change the landscape and carry out construction work. There were prioritized offenses in terms of elements of organized crime in a professional context (Holmøy, 2020a; Wang, 2019; Ytreberg, 2018).

One of the hypotheses by Billiet et al. (2014) addressed potential differences in judgments of justification for punishment between courts of first instance and courts of appeal. The hypothesis suggests that the courts of first instance are more likely to reflect the core goals of environmental law, namely the protection of the environment for health reasons and for the value of nature. Judgments from the courts of appeal, however, would be more likely to reflect the core goals and principles of criminal law, namely the protection of individuals and properties. The researchers could confirm this hypothesis.

Green Criminology Perspectives

Green criminology is concerned with environmental injustice (Lynch & Stretesky, 2013: 9):

> Green criminology was originally conceived as a political economic explanation of how environmental deviance and crime may be influenced by variations in economic interests that are tied to particular types of production. Borrowing from the green social movement and green political movement literatures, the environmental justice literature and political economic theory, green criminology establishes the rational for examining environmental crime, deviance, law, and social control in relation to the political economy of production, and class and race conflict and inequality. In this view, green criminology is similar to radical or critical criminology, but shifts the focus of research to environmental harms and social control.

Class and inequality are mentioned in this quote. This is interesting in the perspective of the Tjøme case that is extensively presented in this book. The rich and mighty belong to the upper-level class, and inequality exists in access to resources to commit and defend deviance. Furthermore, radical or critical criminology derives from a political movement, where social democrats have ruled Norway for most of the years after World War II. Had right-wing rather than left-wing governments ruled Norway for all these years, green criminology would not have had the same impact on policing environmental crime. In fact, Økokrim established to investigate and prosecute financial crime in the elite by white-collar offenders, has recently stepped up its attention toward environmental crime as an extension of traditional white-collar crime. Traditional fraud and corruption are no longer more important than economically motivated and irreversible harm to public health and nature (Barnes & White, 2020; Long et al., 2012; Thomson et al., 2020).

As noted by Lynch and Stretesky (2013: 10), environmental law enforcement has an economic basis inherent in class interests:

> As a result, the economic conflicts involved in the development of informal surveillance may be complex and the inability of environmental agencies to fulfill their missions that results from underfunding and understaffing may occur in a political environment in which funding for environmental enforcement is undermined by corporate lobbying efforts.

This reasoning is in line with the perspective of social conflicts. In addition to being too big to fail and too powerful to jail (Pontell et al., 2014), professional characteristics of offenders are such that they belong to the elite in society that make the laws. In the extreme, elite members when violating the law may change the law. Lobbying efforts were visible in Norway in 2020 to open the 100- meters beach zone for privatization. The social conflict perspective suggests that the powerful and wealthy in the upper class of society define what is right and what is wrong. The rich and mighty people can behave like robber barons because they make the laws and because they control law enforcement. The ruling class does not consider white-collar offenses as regular crime, and certainly not like street crime. When a white-collar offender recently had to go to jail because of financial crime, he said on Norwegian television: "This is a really strange situation for me. I must go to jail. But in jail there are only criminals."

The social conflict perspective views financial crime as a function of the conflict that exists in society. The perspective suggests that class conflict causes crime in any society, and that those in power create laws to protect their rights and interests. For example, embezzlement by employees is as a violation of law to protect the interests of the employer. However, it might be argued that an employer must and should protect own assets. Bank fraud is a crime to protect the powerful banking sector. However, in the perspective of social conflicts one might argue that a bank should have systems making bank fraud impossible. If an employee has no opportunity to commit embezzlement, and if a fraudster has no opportunity to commit bank fraud, then these kinds of financial crime would not occur, and there would be no need to have laws against such offenses. Law enforcement protects powerful companies against counterfeit products, although they should be able to protect themselves by reducing opportunities to produce counterfeit products.

The social conflict perspective holds that laws and law enforcement are used by dominant groups in society to minimize threats to their interests posed by those whom they perceive as dangerous and greedy. Crime is defined by legal codes and sanctioned by institutions of criminal justice to secure order in society. The ruling class secures order in the ruled class by means of laws and law enforcement. Conflicts and clashes between interest groups are restrained and stabilized by law enforcement.

According to the social conflict perspective, the justice system is biased and designed to protect the wealthy and powerful. The wealthy and powerful can take substantial assets out of their own companies at their own discretion whenever they like, although employed workers in the companies were the ones who created the values. The superrich can exploit their own wealth that they created as owners of corporations if they do not hurt other shareholders. Employees have no right to object. It is no crime to take out values from own enterprises and build private mansions for the money. This is no crime by the owners. Even when the owners just inherited the wealth created by earlier generations, they can dispose freely of it for private consumption. Similarly, top executives who are on each other's corporate boards grant each other salaries that are ten or twenty times higher than regular employee salaries. As Haines (2014: 21) puts it, "financial practices that threaten corporate interests, such as embezzlement, are clearly identified as criminal even as obscenely high salaries remain relatively untouched by regulatory controls." Furthermore, sharp practices such as insider trading that threaten confidence in equities markets have enjoyed vigorous prosecution, since the powerful see them as opaque transactions that give an unfair advantage to those who are not members of the market institutions.

Karl Marx, who analyzed capitalism and suggested the transition to socialism and ultimately to communism, created the basis for the social conflict perspective. Capitalism is an economic system in which persons privately own trade, industries, firms, shops, and means of production and operate these enterprises or profit. Socialism is an economic system characterized by cooperative enterprises, common ownership, and state ownership. Communism is a socioeconomic system structured upon the common ownership of the means of production and characterized by the absence of social classes.

Marxist criminology views the competitive nature of the capitalist system as a major cause of financial crime. It focuses on what creates stability and continuity in society, and it adopts a predefined political philosophy. Marxist criminology focuses on why things change by identifying the disruptive forces in capitalist societies, and describing how power, wealth, prestige, and perceptions of the world divide every society. The economic struggle is the central venue for the Marxists. Marx divided society into two unequal classes and demonstrated the inequality in the historical transition from patrician and slave to capitalist and wage worker. It is the rulers versus the ruled. Marx also underlined that all societies have a certain hierarchy wherein the higher class has more privileges than the lower one. In a capitalist society, where economic resources equate to power it is in the interest of the ascendant class to maintain economic stratification in order to dictate the legal order.

Green criminology is partly based on the perspective of social conflicts. As suggested by Lynch (2020), environmental crime has created new forms of conflict, harm, injury, and damage.

Lynch and Stretesky (2013) studied policing of environmental crime. They found that citizens can act as important adjuncts in the enforcement of laws and regulations, and that the increased citizen participation creates growth in environmental policing. Their finding is supported in the case study of Tjøme, where whistleblowers Amundsen and Ravndal provided valuable information to police investigators regarding municipal misconduct and summerhouse owners' wrongdoing on the shoreline.

Green Criminology Politics

As suggested above, green criminology and policing of environmental crime is a matter of politics. In the long run, law enforcement has to stay consistent with the priorities of the citizens in general and the ruling class in particular. Returning now to the shoreline in Norway, it is obviously a matter of politics to what extent environmental wrongdoing in the beach zone is serious. Therefore, it came as no surprise as the next newspaper to pursue the issue of the rich and mighty on the shoreline was the left-wing daily Norwegian newspaper 'Klassekampen' (class struggle). On Saturday, July 4, 2020, the newspaper had on its front page a picture of an untouched landscape where fifty luxury vacation homes will destroy public access, animal life, and natural developments (Tollersrud, 2020: 1):

> Bit by bit, the country is being built down. On a green promontory on Kvalnes in Bodø, 50 planned cabins have split the local community.

The local community was split, because some were concerned with environmental protection, while others were concerned with depopulation and jobs. The area had for a long time been depopulated by young people moving into cities for education and work. Vacation homes would provide jobs in both the construction periods and later service businesses such as caretaking of boats, maintenance of houses, food stores, and other service facilities. The skeptics do not believe in it (Tollersrud, 2020: 27):

> They do not believe that the cabin people will have positive impacts on the rural community. They do not know of any jobs associated with the cabins located here already. The impression is that the cabin people shop at the major grocery stores in the city on their way out and use the local shop only if there is something they forgot to buy. They do not use the bus and do not go to school here.
>
> -When people come to the cabin, I think they will be at peace. They rarely become part of the village. If I need milk, I won't run to the cabin area. There is no one there, or I may not know them. It's not the same as having neighbors, that you see every day in the rain and sun, Monday, Tuesday, and Wednesday. Without the villagers, the village ravages, Bjune says.

The politics of green criminology and environmental crime policing was also visible when the Parliamentary Ombudsman—the ombudsman appointed by the Norwegian Parliament to safeguard the rights of individual citizens in their dealings with the public administration—found that municipal permits violate the rights of individual citizens. Municipal permits for mansion construction violate the rights of individual citizens in Norway to move freely along the coast.

Around seventy people work at the Parliamentary Ombudsman's office, around forty of which are legal advisers (www.sivilombudsmannen.no). The organization comprises three departments that consider complaint cases from persons who believe they have suffered an injustice on the part of the public administration, as well as an investigation department that raise matters on its own initiative. The investigation into summerhouses was raised as a matter on the ombudsman's own initiative.

The interesting aspect of the investigation by the ombudsman was the selection of three municipalities in Norway that had a sensationally large number of permits that the municipalities had labeled exemptions. The cases were exemptions to the prohibition of buildings in the beach zone. Municipalities in Norway granted exemptions in 751 out of 1319 application cases for new buildings in the beach zone in 2018. The ombudsman selected Kragerø municipality with 563 exemptions, Mandal municipality with 181 exemptions, and Askøy municipality with 121 exemptions. Kragerø municipality is almost as attractive as Færder municipality (previously Tjøme municipality) in terms of vacation resorts for the rich and powerful. While the climate is about the same, Færder is less than two hours drive away from Oslo, while Kragerø is a three-hour drive from the capital of Norway.

In their report on Kragerø, the Ombudsman (2020: 15) draws the following conclusion:

- The majority of decisions is deficient and leaves doubt as to whether the municipality has made a real advantage/disadvantage assessment.
- The majority of decisions lack description for relevant benefits, as required by law.
- Many decisions are justified by circumstances that are not relevant or have little weight in the advantage/disadvantage assessment.
- There is doubt whether the municipality has adequately reviewed and described any disadvantages of the exemptions.

Getting an exemption from the building ban in the beach zone is positive for the individual developer. However, if exemptions are granted in violation of the Planning and Building Act, it may constitute an injustice to other citizens, whether they are neighbors or part of a wider circle. The Parliamentary Ombudsman shall seek to ensure that the public administration does not practice injustice against individual citizens. Examining how the public administration carries out its tasks in which the interests of different citizens oppose each other is part of the mandate. Where there is any contradiction between a mansion builder and the public interest, there will often be no natural advocate for the public interest in a specific case. Then the ombudsman becomes an important agency to examine on behalf of the public interest.

The powers of the Parliamentary Ombudsman are limited. There is room for evidence collection and conclusions as a supplement to regular courts, but the ombudsman cannot pass verdicts on individuals or institutions. However, since the office of the ombudsman was created in 1962, most public authorities have adapted recommendations from the ombudsman after investigations took place.

Kragerø municipality approved three out of four applications for exemptions that are 75 percent, while Mandal municipality approved 96 percent (Klevstrand & Espeland, 2020: 15):

> The high number of exemptions indicates that it must be tightened. As many as 98 percent of all cases have incomplete reasons that we cannot accept. Here, the municipal politicians must understand that they too carry responsibility to preserve nature, both for the nature itself and the public. This is a responsibility they take on terribly light, says Silje Lundberg in the Conservation Federation.

Green Criminology Research

Environmental crime is a complex and ambiguous term often used to describe transnational scandals where prominent politicians and corporations cause climate change, species loss, and human suffering (Lynch, 2020). In this perspective, the mansions on the shoreline in Norway can seem rather insignificant as environmental crime. However, as argued by the slippery slope perspective (Arjoon, 2008: 78), "it is the small infractions that can lead to the larger ones." Sliding on the slippery slope can make the offender and society feel innocent, as people did not notice ending up on the wrong side of environmental protection (Welsh et al., 2014). A series of small infractions gradually increase over time. Committing small indiscretions over time gradually lead people to complete larger unethical acts that they otherwise would have judged impermissible (Murphy & Dacin, 2011; Pettigrew, 2018).

Gibbs and Boratto (2017) suggest that environmental crime is sometimes used as an umbrella term for crime related to biodiversity, wildlife, animals, natural resources, hazardous waste, banned substances, and environmental quality. Johnson et al. (2020) found that the perspective of criminalization of all kinds of environmental crime has changed from civil and administrative solutions to public criminal prosecution.

Lynch (2020: 50) observed three decades of green criminology (GC) research:

> Three decades after its introduction, the scope of GC is now broad. It is not simply about ecological harms (e.g., pollution) or explaining the origins of those harms. It now also draws attention to the construction, implementation, and enforcement of environmental laws; to the problem of environmental justice; to the ways in which all living beings are victimized by ecological harms. And, like any other area of criminological research, GC must also address the definition of its main focus, green crimes, which is not a simple nor settled matter.

Lynch (2020) expresses five challenges in green criminology research. First, neither green criminology nor environmental crime is well-defined concepts. At its most simple level, the concepts are defined in terms of a series of examples. A well-defined concept needs to define what it is and what it is not. A concept in criminology can be defined by the offender, the offense, or the victim. It seems that applied definitions typically include some description of the offense and some description of whom or what is harmed.

The second challenge is measuring green crime. Measuring is concerned with both the seriousness of each incident and the frequency of incidents. The seriousness of white-collar crime is typically measured in the monetary equivalent of the illegitimate gain or the illegitimate savings in costs. The frequency or magnitude of white-collar crime is typically measured in terms of detected versus undetected offenses, where a typical estimate is that one out of ten, or one out of twenty offenses are detected and prosecuted. However, putting a monetary value of environmental harms is not a simple task, and counting the incidents of green crime is challenging as well.

The third challenge in green criminology research is addressing legal statutes and law enforcement mechanisms that define and control environmental crime (Lynch, 2020: 51):

> Unlike street crime – but like corporate and white-collar crime – green crimes are defined in administrative, regulatory, and civil laws, as well as criminal laws, and at various levels of governance. Consequently, the traditional legal definition of crime as a violation of criminal law does not strictly apply to green crimes.

This challenge is evident in the presented Tjøme case. The employees in the municipality granting permits and the architect applying for permits from the municipality have a traditional relationship that can be addressed by the traditional legal definition of corruption. That is simple for the first round of prosecution in January 2021. Prosecuting the summerhouse owners, however, is a more complicated matter in terms of administrative, regulatory, civil laws, as well as criminal laws that may or may not apply. A complicating issue for research in green criminology as well as identification of laws and rules that officially describe environmental crime, is the often massive and complex documents on the topic that have all kinds of links between each other. Mansions owners can argue that rule complexity has prevented compliance (Lehman et al., 2019). Green crime is a social construction (Lynch, 2020: 51):

> As a social construction, environmental laws and regulations involve conflicts about representing scientific knowledge, corporate interests, and victims' interests (i.e., any living being's interests) in those rules. The outcome often represents the interests of powerful groups and the structural economic organization of society rather than scientific knowledge.

Summerhouse owners in the beach zone on the island of Tjøme may certainly argue that the tremendously long coastline of Norway has plenty of shorelines totally untouched. They may argue that their own small spots represent a valuable recreational opportunity for their families and no harm to anyone else. On the other hand, powerful environmental movements with their lobbying skills may find prestige in preventing anyone from changing the landscape anywhere. The outcome thus can result from the relative powers of interest groups (Lynch, 2020: 51):

> By moving beyond the law, GC recognizes and gives voice to a host of victims (human, nonhuman, and ecosystem/ecological) that traditional criminology willingly ignores and abandons, making green victimization a fourth area of concern.

The fourth challenge in green criminology research is thus green victims that come in many forms. In the Vest Tank case, local residents were victims of pollutants. In the Tjøme case, victims were tourists and other visitors denied access to recreational areas, as well as not so fortunate cabin owners on the island without own access to beach areas. The less fortunate cabin owners may also have experienced unequal enforcement of environmental laws on the island, which is another kind of victimization in green criminology. Some of these victims have no voice and, to be protected, require representation by police agencies such as Økokrim and political agencies such as the Parliamentary Ombudsman.

The fifth and final concern in green criminology research is the dominance of quantitative studies in criminology with a fundamental bias toward street crime. Environmental crime research has long been descriptive, theoretical, philosophical, and qualitative. Both the Vest Tank case and the Tjøme case are typical examples. Lynch (2020) argues that green criminology requires more extensive quantitative research to legitimize the importance of environmental crime studies.

While environmental crime is a criminological matter, green criminology research reflects intellectual traditions from numerous disciplines including social conflict as defined in Marxist analysis (Lynch, 2020: 52):

> GC reflects intellectual traditions from numerous disciplines and would be impossible without Marxist analysis, scientific risk assessment, environmental sociology, animal rights studies, environmental justice research, ecophilosophy, and so on.

Research needs to address how environmental crime is conceptualized and how criminalization processes develop. Certain mechanisms are interesting to study, which cause some of the harms from environmental wrongdoing to be legal while others are criminalized. Law enforcement and punishment, or lack of punishment, in relation to the extent of harm need more research, as well as how harms should be understood and how they should be addressed. Moral rights and perspectives of justice influence what is considered right and wrong to do.

As mentioned in the introduction to this chapter, green criminology has become an important and expanding field in criminology, particularly within the perspective of critical criminology, which focuses on challenging traditional understandings and uncovering false beliefs about criminal justice. Expansion in the field of green criminology research is taking place in several directions. An example is the study of green cultural criminology by Redmon (2018), which explores how environmental crime is perceived culturally. Green cultural criminology has examined how media represents harm against species, ecology, and humankind. Such examinations show that representations and perceptions of environmental crime influence the extent to which people experience affective engagement with justice and injustice. Generally, green cultural criminology questions taken-for-granted assumptions about offending and harm.

Another example is the study of wildlife crime research by McFann and Pires (2020), who found that green criminologists seldom conduct quantitative studies. Rather, the majority of investigations were conceptual or theoretical in nature. The study shows a Western dominance in wildlife crime research while the majority of the high-profile wildlife crime emanates from non-Western institutions. This kind of study is very much in line with the critical criminology perspective among green criminologists.

Bibliography

Agder. (2019, September 16). Court case 18-178766AST-ALAG. *Agder lagmannsrett (Agder court of appeals)*, Agder, Norway, 22 pp.

Arjoon, S. (2008, Spring). Slippery when wet: The real risk in business. *Journal of Markets & Morality, 11*(1), 77–91.

Barnes, A., & White, R. (2020). Mapping emotions: Exploring the impact of the Aussie Farm map. *Journal of Contemporary Criminal Justice, 36*(3), 303–326.

Berglihn, H. (2020, June 29). Hanne Madsen må i fengsel: -Viktig at det reageres strengt (Hanne Madsen must go to jail: -It is important that the reaction is strict). Daily Norwegian business newspaper *Dagens Næringsliv*. www.dn.no.

Billiet, C. M., Blondiau, T., & Rousseau, S. (2014). Punishing environmental crimes: An empirical study from lower courts to the court of appeal. *Regualtion & Governance, 8*, 472–496.

DN. (2019, September 21). Ny dom kan skremme pengesterke hytteeiere fra ulovlig bygging – Ny dom gir god hjelp til kommuner som vil stoppe ulovlige byggearbeider i strandsonen (New court sentence might scare money-rich cottage owners from illegal construction building—New court sentence provides good help to municipalities who want to stop illegal construction work in the beach zone). Daily Norwegian business newspaper *Dagens Næringsliv*, p. 2.

Gibbs, C., & Boratto, R. (2017). Environmental crime. In H. N. Pontell (Ed.), *Oxford encyclopedia of criminology and criminal justice*, published online https://oxfordre.com/criminology.

Haines, F. (2014). Corporate fraud as misplaced confidence? Exploring Ambiguity in the accuracy of accounts and the materiality of money. *Theoretical Criminology, 18*(1), 20–37.

Holmøy, E. (2020a, June 25). Ny anmeldelse på vei til Økokrim: -Vi har levert omfattende dokumentasjon (New report on its way to, Økokrim: -We have delivered comprehensive documentation). Local daily newspaper *Tønsbergs Blad*. www.tb.no.

Holmøy, E. (2020b, October 16). Advokat Anne Tellefsen klaget helt opp til departementet for at hennes millionærklient skal få dekket utgiftene sine. Nå har svaret kommet (Lawyer Anne Tellefsen complained all the way to the ministryt hat her millionaire client should have her expenses covered. Now the answer has come). Local daily newspaper *Tønsbergs Blad*. www.tb.no.

Johnson, E. W., Schwartz, J., & Inlow, A. R. (2020). The criminalization of environmental harm: A study of the most serious environmental offenses prosecuted by the U.S. federal government, 1985–2010. *Environmental Sociology*, published online https://doi.org/10.1080/23251042.2020.1748269.

Klevstrand, A., & Espeland, J. M. (2020, July 4). Sa ja til 96 prosent av strandsonesøknadene (Said yes to 96 percent of the beach zone applications). Norwegian daily business newspaper *Dagens Næringsliv*, pp. 14–15.

Lehman, D. W., Cooil, B., & Ramanujam, R. (2019). The effects of rule complexity on organizational noncompliance and remediation: Evidence from restaurant health inspections. *Journal of Management*, 1–33, published online, https://doi.org/10.1177/0149206319842262.

Long, M. A., Stretesky, P. B., Lynch, M. J., & Fenwick, E. (2012). Crime in the coal industry: Implications for green criminology and treadmill of production. *Organization & Environment, 25*(3), 328–346.

Lynch, M. J. (2020). Green criminology and environmental crime: Criminology that matters in the age of global ecological collapse. *Journal of White Collar and Corporate Crime, 1*(1), 50–61.

Lynch, M. J., & Stretesky, P. B. (2013). The distribution of water-monitoring organizations across states: Implications for community environmental policing and social justice. *Policing: An International Journal of Police Strategies & Management, 36*(1), 6–26.

McFann, S. C., & Pires, S. F. (2020). Taking stock in wildlife crime research: Trends and implications for future research. *Deviant Behavior, 41*(1), 118–135.

Murphy, P. R., & Dacin, M. T. (2011). Psychological pathways to fraud: Understanding and preventing fraud in organizations. *Journal of Business Ethics, 101*, 601–618.

Ombudsman. (2020). *Dispensasjoner i sandsonen (Exemptions in the beach zone).* Sivilombudsmannen (Parliamentary Ombudsman), Oslo, Norway.

Pettigrew, W. A. (2018). The changing place of fraud in seventeenth-century public debates about international trading corporations. *Business History, 60*(3), 305–320.

Pontell, H. N., Black, W. K., & Geis, G. (2014). Too big to fail, too powerful to jail? On the absence of criminal prosecutions after the 2008 financial meltdown. *Crime, Law and Social Change, 61*(1), 1–13.

Redmon, D. (2018). Video methods, green cultural criminology, and the anthropocene: Sanctuary as a case study. *Deviant Behavior, 39*(4), 495–511.

Supreme Court. (2020, June 26). Case HR-2020-1353-A (sak nr. 19-158426STR-HRET). *Norges Høyesterett (Norwegian Supreme Court)*, Oslo, Norway.

Thomson, R., Espin, J., & Samuels-Jones, T. (2020). Green crime havens: A spatial cluster analysis of environmental crime. *Social Science Quarterly, 101*(2), 503–513.

Tollersrud, E. (2020, July 4). Bygda som vil level (The village that will live). Daily Norwegian newspaper *Klassekampen*, pp. 1 and 18–27.
Wang, M. (2019, December 5). Mer trøbbel for hyttemillionær. Kommunen mener alle gamle tillatelser er ugyldige (More trouble for cabin millionaire: The municipality says that all old permits are invalid). Norwegian local newspaper *Tønsbergs Blad*. www.tb.no.
Welsh, D. T., Ordonez, L. D., Snyder, D. G., & Christian, M. S. (2014). The slippery slope: How small ethical transgressions pave the way for larger future transgressions. *Journal of Applied Psychology, 100*(1), 114–127.
Ytreberg, R. (2018, April 30). Kommunen stoppet Breili-prosjekt på Tjøme (The municipality stops Breili project on Tjøme). Norwegian daily business newspaper *Dagens Næringsliv*, pp. 4–5.

11

Private Internal Policing

Private policing of white-collar crime has been a controversial issue for quite some time. Fraud examiners from global auditing firms and local law firms conduct internal investigations, resulting in reports of investigations that are the clients' confidential property. This chapter presents four disclosed reports of investigations to illustrate the theory of convenience for white-collar crime and to determine the examination maturity in private policing. The suggested maturity model with five levels determines whether the investigation as chaotic or messy, or whether the investigation contributed to disclosure or clarification. The most mature level is investment, where the benefits of the examination exceed the costs of the examination. Many fraud examiners have a long way to go before they deserve the label of professional investigators or examiners. Future research may help professionalize the business of private policing of economic crime. The four reports of investigations are from Moldova, the Netherlands, Sweden, and Switzerland (Gottschalk, 2020). None of them are concerned with environmental crime, as no such reports have been possible to identify and retrieve.

Two decades have soon passed since Schneider (2006) and Williams (2005a, 2005b) emphasized the problematic role of fraud examiners

P. Gottschalk, *Detecting and Investigating Environmental Crime*,
https://doi.org/10.1007/978-3-030-74184-6_11

in internal investigations, as examiners conduct private policing of economic crime. In the meantime, the business of fraud examinations has grown substantially in forensic departments in auditing firms and in consulting departments in law firms. Examples include Clifford Chance (2020), Garcia (2014), Kroll (2016, 2017), and Sands (2019a, 2019b). This chapter develops these four examples into case studies of private policing of economic crime in terms of internal investigations by fraud examiners. The purpose is to illustrate various levels of maturity in the private policing business.

Fraud Investigation Reports

Private policing is provision of policing services such as investigations and detective work other than by public servants in the normal course of their public duties. The public policing function is typically located in a government body, while private policing in the area of fraud examinations is typically located in auditing firms and law firms. The authority and tools of a policing entity ranges from full legal authority granted to the office of the public police, to no special authority at all beyond that provided by the clients of fraud examinations. Beneficiaries can range from everyone as in the case of public policing, to only those entities that fund policing. Furthermore, "the public police, as an institution of the criminal justice system, inherently possess a punishment or coercion-based mentality propelled by their authority as agents of the criminal law and their capacity to apply state-sanctioned force" (Wood, 2020: 25). In contrast, private policing of economic crime is consulting work where fraud examiners conduct internal investigations to establish facts and advise clients what to do next.

The business of private policing by fraud examiners seems to be growing more rapidly than national and international public police organizations to combat fraud (Wood, 2020: 26):

> Determining the ratios of private agents to public agents is invariability hard to do with much accuracy, but it is clear that private policing entities surpass the latter in sheer numbers and to varying degrees across established democracies and countries in transition.

The purpose of this article is to study cases of internal investigations by fraud examiners. The article presents a stage model for maturity in private policing and applies the maturity model to each case study. The four cases resulted from a search for investigation reports. Client firms who hire investigators tend to keep internal reports by fraud examiners secret. They do not want to disclose the contents to the media or the public in general. Even when there is suspicion of serious economic crime, many clients prefer not to disclose reports to the police (Gottschalk & Tcherni-Buzzeo, 2017). "Typically, corporations tend to take their own measures to resolve instances of fraud without resorting to the police" (King, 2020a, 2020b). Therefore, we cannot generalize the extent of fraud examination maturity based on our sample of only four reports. King (2020a) argues that nine attributes are critical for the corporate investigator's inquiry success: communication, motivation, industry experience, qualifications, police or regulatory experience, business acumen, conceptual thinking, resilience, and rapport building.

Figure 11.1 illustrates the maturity model that we apply in this research. Stages of growth models for maturity levels help to assess and evaluate a variety of phenomena (e.g., Röglinger et al., 2012; Solli-Sæther & Gottschalk, 2015). Stage models predict the development or evolution of investigative maturity from basic performance to superior results (Iannacci et al., 2019: 310):

> They also suggest that this development is progressive (i.e., each successive stage is better than the previous one), stepwise (i.e., each step is a necessary prerequisite for the following step in the sequence), and prescriptive (i.e., each step must occur in a prescribed order in accordance with a pre-existing plan or vision), thus emphasizing the chain of successful events rather than the mechanisms by which subsequent stages come about.

Here we apply the concept of maturity levels to evaluate private internal investigations (Brooks & Button, 2011; Button & Gee, 2013; Button

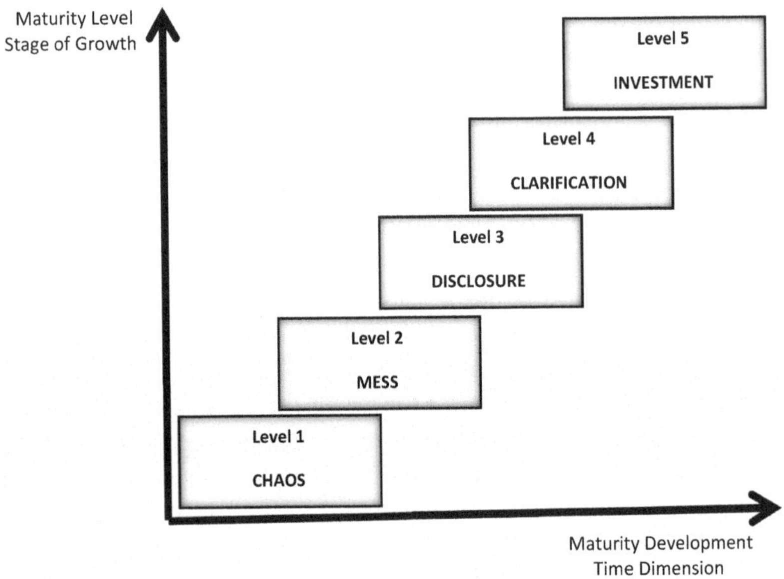

Fig. 11.1 Maturity model for internal private investigations with five stages

et al., 2007a, 2007b; Schneider, 2006; Williams, 2005a, 2005b). The purpose is to develop characteristics of investigations at different maturity levels. Based on previous studies of investigation reports (Gottschalk, 2020), we present a five-stage model as illustrated in the figure.

The five maturity levels in the stage model for private policing have the following descriptions:

1. *The investigation was a chaos*. The investigation caused more confusion than before the examination was initiated. The investigation was insufficient, inadequate, surface-oriented, a waste of time, useless, passive, unprofessional, worthless, immature, unacceptable, bad, meaningless, fruitless, awful, and chaotic. The investigation was a failure and a disaster.
2. *The investigation was a mess*. Nothing came out of the investigation. The investigation was random, amateur, formalities focused, somewhat good, sufficient, descriptive, problem-oriented, neutral, unsystematic, inadequate, activity-oriented, shortsighted, fruitless,

deviations-oriented, reactive, questions-oriented, and messy. The investigation lacked scrutiny, was a collection of information without analysis, and had too many assumptions that made conclusions invalid.

3. *The investigation was a disclosure.* Examiners were successful in identifying and documenting some new facts. The investigation had a clear perspective; it was competence oriented, average, biased, targeted, systematized, integrated, moderate, indifferent, standard, competent, cause-based, revealing, and disclosure-oriented. The investigation was problem-oriented and limited by the mandate.
4. *The investigation was a clarification.* The investigation was able to reconstruct past events and sequences of events. The investigation was responsible, detailed, conscientious, sufficient, professional, neutral, unprejudiced, integrated, proactive, preventive, mature, competent, systematic, professional, explorative, immaculate, expedient, truth seeking, facts-based, complete, independent, and clarifying. The investigation added value.
5. *The investigation was an investment.* The investigation made a valuable contribution to the organization, where investigation benefits exceed investigation costs. The investigation was optimal, innovative, profitable, strategic, extraordinary, outstanding, provident, value-oriented, advanced, learning-focused, valuable, irreversible, truth-based, socially responsible, exceptional, excellent, perfect, exemplary, and a profitable investment. The investigation was a masterpiece and enrichment for the client and society.

Each case study in this chapter applies the described model of maturity in private policing. Each case study also applies the structural model of convenience theory as illustrated in Fig. 2.1. The emerging theory of convenience for white-collar crime is concerned with financial possibilities and threats, organizational opportunity to commit and conceal financial crime, as well as personal willingness for deviant behavior. The research literature on convenience theory is growing (e.g., Braaten & Vaughn, 2019; Chan & Gibbs, 2020; Gottschalk, 2019; Hansen, 2020; Kireenko et al., 2019; Leasure & Zhang, 2018; Otu & Okon, 2019; Reese & McDougal, 2018; Vasiu & Podgor, 2019).

White-collar offenders commit economic crime where a great variety of options of financial misconduct is available to them, as illustrated in Fig. 11.2. Fraud, theft, manipulation, and corruption are four main categories of financial crime with a number of sub-categories. Fraud is intentional perversion of the truth for inducing another in reliance upon

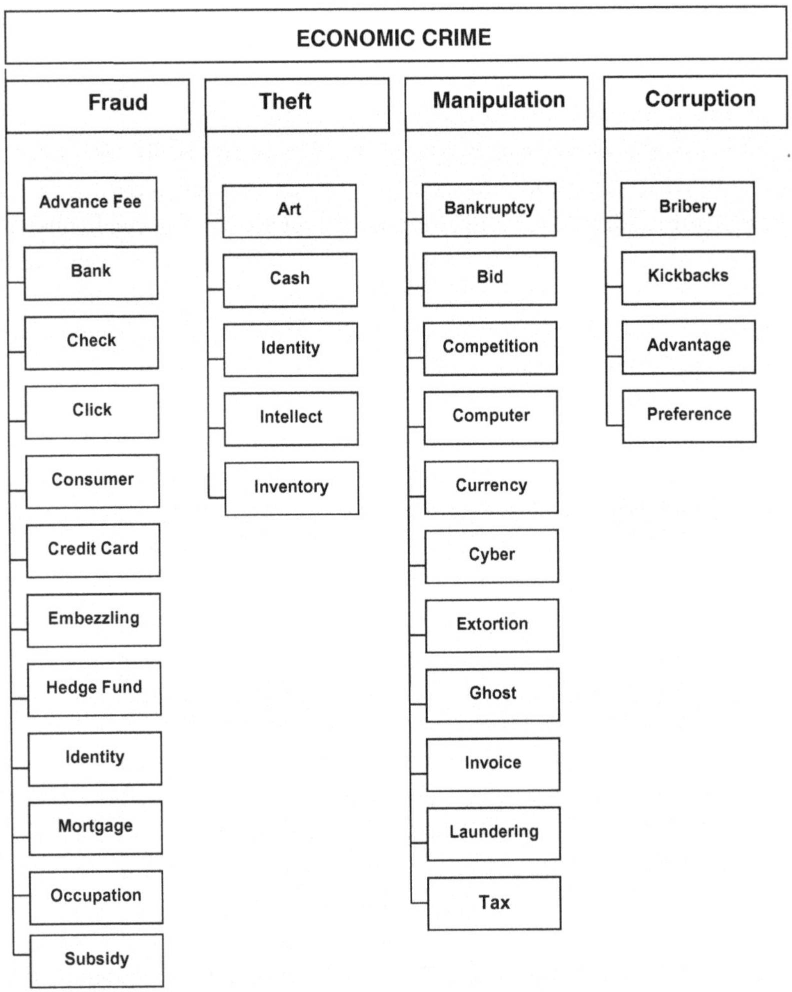

Fig. 11.2 Main categories and sub-categories of e-crime

it to part with some valuable thing belonging to him or to surrender a legal right. Fraud is unlawful and intentional making of a misrepresentation, which causes actual prejudice or which is potentially prejudicial to another. Bank fraud is a typical example. Bank fraud is a criminal offence of knowingly executing a scheme to defraud a financial institution.

Theft is the illegal taking of property from another person, group, or organization without the victim's consent. For example, identity theft combined with identity fraud is the unlawful use of another's personal identifying information. It involves financial or other personal information stolen with the intent of establishing another person's identity as the thief's own. It occurs when someone uses personally identifying information, like name, social security number and date of birth, government passport number or credit card number without the owners' permission, to commit economic crime.

Manipulation is a form of cheating where the offender breaks the rules while deliberately leading or allowing others to think the offender respects the rules (Eabrasu, 2020). Manipulation is a means of gaining illegal control or influence over others' activities, tools, and results. For example, bankruptcy crime is criminal acts committed in connection with bankruptcy or liquidation proceedings. A person filing for bankruptcy or a business that has gone into liquidation can hide assets after proceedings start. Thereby fraudsters can prevent creditors from collecting their valuable belongings. However, most of the criminal acts are typically committed before bankruptcy/liquidation proceedings start, e.g., the debtor has failed to keep accounts or has unlawfully withdrawn money from the business.

Corruption is the giving, requesting, receiving, or accepting of an improper advantage related to a position, office, or assignment. The improper advantage does not have to relate to a specific action or to not doing this action. It will be sufficient if the advantage connects to a person's position, office, or assignment. An individual or group is guilty of corruption if they accept money or money's worth for doing something that he is under a duty to do anyway, that he is under a duty not to do, or to exercise a legitimate discretion for improper reason. Corruption is to destroy or pervert the integrity or fidelity of a person in his discharge of duty, it is to induce to act dishonestly or unfaithfully, it is to make venal, and it is to bribe. Corruption involves behavior on the part of officials in the public or private sectors, in which they improperly and unlawfully enrich themselves and/or those close to them, or induce others to do so, by misusing the position in which they find themselves. Corruption covers a wide range of illegal activity such as kickbacks, embezzlement, and extortion.

The methodology applied to investigation reports in this research is content analysis (Bell et al., 2018; Braaten & Vaughn, 2019; Saunders et al., 2007). Content analysis is any methodology or procedure that works to identify characteristics within texts attempting to make valid inferences (Krippendorff, 1980; Patrucco et al., 2017). Content analysis assumes that language reflects both how people understand their surroundings and their cognitive processes. Therefore, content analysis makes it possible to identify and determine relevant text in a context (McClelland et al., 2010).

Swedbank by Clifford Chance

Swedbank carried out bank transactions of more than 37 billion Euros (about US$40 billion) with a high risk for money laundering over a five-year period according to private policing in terms of an internal investigation by fraud examiners from law firm Clifford Chance (2020). The investigation report suggests that the Swedish bank actively targeted high-risk individuals in the Baltic region and points to failings from both top management and the board (Milne, 2020).

Birgitte Bonnessen was the chief executive officer (CEO) at Swedbank. She had to leave the position in 2019 (Makortoff, 2019). When Clifford Chance (2020) presented their report of investigation, the new Swedbank board decided to withdraw her final compensation of 26 million Swedish kroner (US$2.7 million). At the same time, Swedbank accepted a fine of 4 billion Swedish kroner (US$408 million) from the Swedish finance inspection.

White-Collar Convenience

The bank benefitted from having oligarchs and other Russians as customers without implementing anti-laundering procedures. The motive for violating anti-money laundering laws was for Swedbank to make extraordinary profits by corporate crime (Naylor, 2003), as illustrated by the convenience theme motive-possibilities-corporate in Fig. 11.3.

Swedbank executives had a high status (Agnew, 2014), that allowed them to implement special arrangements for non-resident clients. They could claim that the act of wrongdoing is morally justifiable (Schnatterly

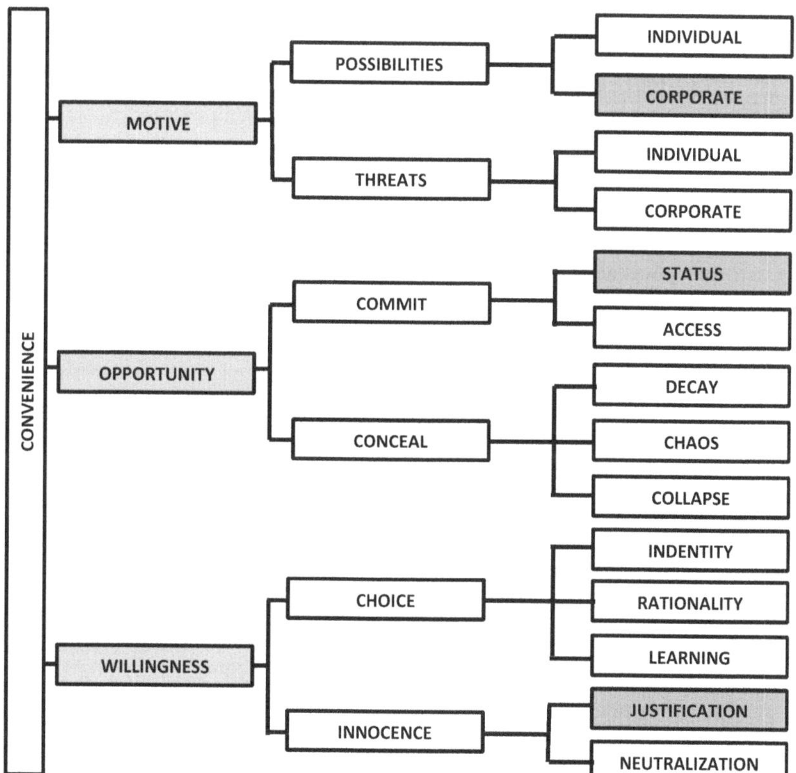

Fig. 11.3 Convenience themes in the case of Swedbank

et al., 2018), probably based on upper echelon information selection (Gamache & McNamara, 2019). A CEO such as Birgitte Bonnessen might suffer from narcissistic traits (Zhu & Chen, 2015), where she constructed her own professional world (Chatterjee & Pollock, 2017). Many CEOs tend to have a temporal focus (Gamache & McNamara, 2019). Schnatterly et al. (2018) found that pressure, opportunity, and rationalization cause CEO wrongdoing.

Fraud Examination Outcome

Fraud examiners from law firm Clifford Chance failed in determining whether money laundering had actually occurred in Swedbank's subsidiaries in Estonia, Latvia, or Lithuania. In fact, they did not even try to collect evidence of money laundering activities (Clifford Chance, 2020: 8):

> Inclusion in the group of AML risk identified customers is not necessarily indicative of suspicious behavior or improper conduct. Rather, these parameters were designed to capture the portion of the Baltic subsidiaries' customer population that warranted further review. In addition, the fact that an external payment to or from one of these customers hit against one or more of the detection algorithms does not mean that the payment should have been considered suspicious at the time, let alone is it evidence that a customer engaged in money laundering or other financial crime. Rather the detection algorithms are designed to identify transactions with risk indicators similar to those that a transaction monitoring system would flag for further review.

Rather than collecting evidence on money laundering activities, such as circular transfers by suspected non-resident companies, fraud examiners collected only evidence on misconduct in terms of failing anti-money laundering work in the Baltic subsidiaries. The word "evidence" is subject to repetition several times in the investigation report, however mainly in terms of not identifying evidence rather than identifying evidence. For example, "the investigation did not identify evidence that these issues were presented to the Board" (Clifford Chance, 2020: 83), which simply means that to the extent board members learned about misconduct, this was of course not documented in formal ways. Clifford Chance (2020: 84) continues to emphasize that reports were not "provided to the Board or its committees."

Investigators found evidence of two Russian oligarchs who moved "funds among related accounts in Swedbank Estonia" (Clifford Chance, 2020: 87). Again, investigators failed in identifying potential evidence of money laundering activities among Russian oligarchs and other non-resident bank clients in Estonia, Latvia, or Lithuania.

FTI Consulting provided forensic support to Clifford Chance. They collected and analyzed transactions processed by Swedbank Estonia, Latvia, and Lithuania. In Estonia, FTI identified 338,000 messages that reflected search terms. FTI then determined that 7,100 of these messages contained true hits against search terms defined by the United States Treasury Department's Office of Foreign Assets Control. From further study of those messages, FTI assembled 3,037 transactions for legal review, supplemented by customers from Internet addresses in embargoed countries. The fraud examiners ended up identifying nineteen sanctioned corporate customers. The fraud examiners obtained similar results for Latvia and Lithuania.

Private Policing Evaluation

Chapter VII in the report is presenting the investigation's findings and conclusions. The chapter is 125 pages out of the 216-pages report. The chapter tells a story of developments, but it does not reconstruct past events or sequences of events as required from a fraud examination. Even worse, there are no real conclusions, although the title of Chapter VII claims there will be a conclusion. As a result, and because of the lack of evidence of money laundering, Fig. 11.4 illustrates that the Clifford Chance (2020) investigation report only reaches level two in terms of fraud examination maturity.

Rather than trying to find the underlying cause of rumors and accusations concerning misconduct and crime in the form of money laundering, examiners spend time telling how Swedbank's current board

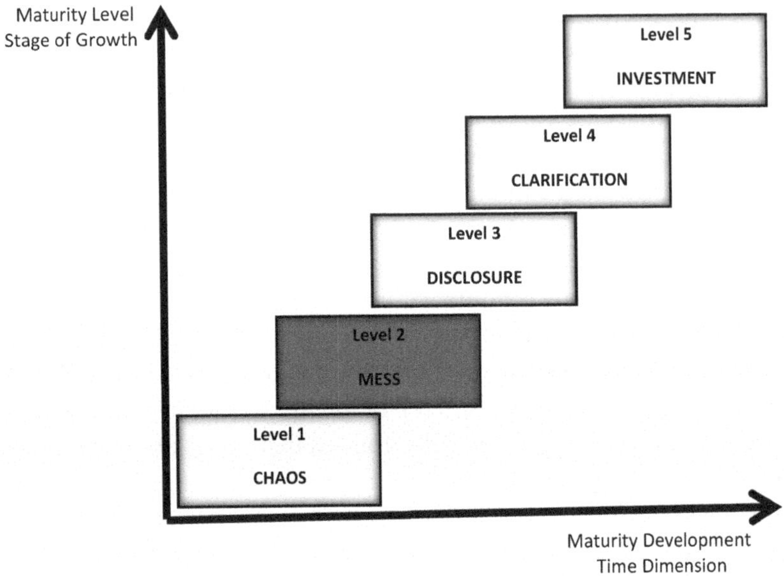

Fig. 11.4 Maturity model for internal private investigations applied to Clifford Chance

and management intend to implement new approaches to compliance and anti-money laundering (Clifford Chance, 2020: 178):

> Swedbank has focused over the past year on transforming its approach to AML/CTF and sanctions compliance, creating new roles, appointing new personnel, increasing resources, revising and strengthening policies and procedures, and taking steps to de-risk its customer portfolio including in the Baltic subsidiaries.

It is naïve to believe that things will change simply because there are changes in top management and because those new executives in charge express ethical intentions. Fraud examiner David di Bari chose to include a whole chapter on remediation and nice words about the current bank management. Maybe he did so to please the new top executives, who are to pay for the investigation by Clifford Chance (2020).

Fifa World Cup by Garcia

Media reports have suggested that the reason why Fifa world cups ended up in Russia in 2018 and in Qatar in 2022 was corruption. Members of the executive committee at Fifa did not only receive bribes from the winning nations Russia and Qatar, but also from losing competitors such as Australia. On December 2, 2010, the executive committee of the international football association Fifa, using an anonymous voting procedure, determined the hosts for the 2018 and 2022 Fifa world cup tournaments. Allegations of corruption related to the voting process had surfaced even before the final vote that December day in Zürich in Switzerland. Ever since, there have been persistent allegations of misconduct with respect to the selection process. In 2017, British newspaper The Guardian reported that a Fifa official allegedly took bribes to back Qatar's 2022 world cup bid (Laughland, 2017):

> Julio Grondona, a senior vice-president at Fifa and head of the Argentinian football association until his death in 2014, allegedly told the witness, Alejandro Burzaco, an Argentinian sports marketing executive, that he was owed the money in exchange for his vote, which helped Qatar secure the lucrative tournament.

Three years earlier, in 2014, the Garcia report was completed. The Garcia (2014) report was an investigation produced by Michael Garcia and Cornel Borbély into allegations of corruption in world association football. Fifa appointed Garcia and Borbély in 2012 to investigate ethical breaches at Fifa, which is world football's governing body. The two examiners quickly focused on persistent public accusations of bribery in the 2018 and 2022 world cup bids, which Russia and Qatar, respectively, won in 2010. Fifa kept the Garcia report secret for several years before it leaked to a German newspaper, which caused Fifa subsequently to release the report to the public by Fifa in 2017.

White-Collar Convenience

Both briber and bribed in a corruption relationship have their financial motives, organizational opportunities, and willingness for deviant behaviors. In Fig. 11.5, illustrations for the briber are with texture, while illustrations for the bribed are with dark grey. The briber has a motive of winning the bid for the world cup, which leads to prestige for the country. It is a situation where business objectives justify means (Jonnergård et al., 2010).

The bribed committee member can make an extraordinary profit, which is possible because of the position at Fifa (Pontell et al., 2014).

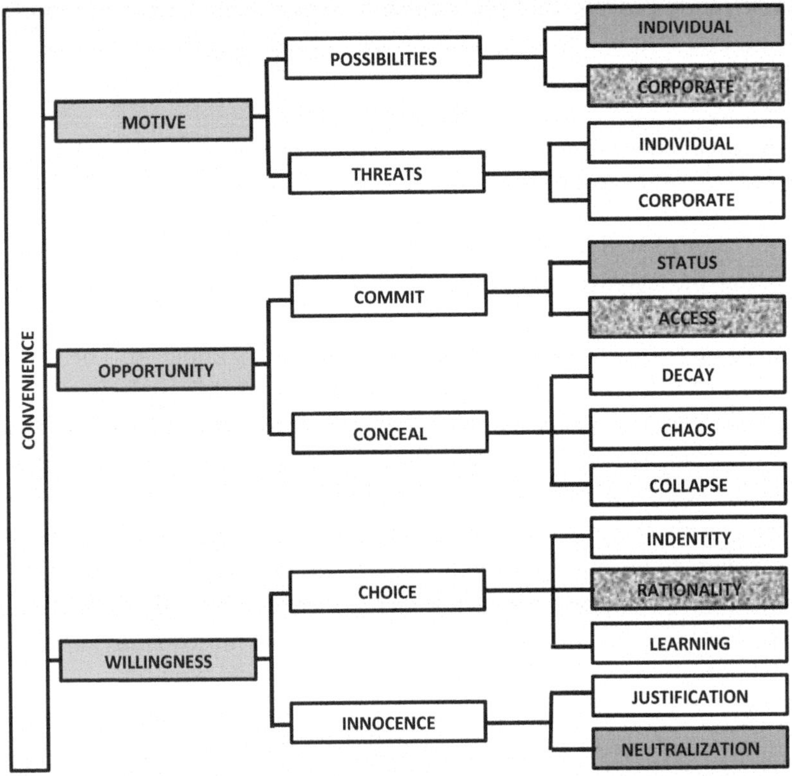

Fig. 11.5 Convenience themes in the case of Fifa

The briber has access to resources to bribe members (Benson & Simpson, 2018). It is a rational choice for the briber to get involved in corruption to improve the nation's chance of winning the bid (Pratt & Cullen, 2005). The bribed can neutralize potential feeling of guilt by arguing that his or her vote did not change (Sykes & Matza, 1957).

Fraud Examination Outcome

In the perspective of evolving accounts as scandals arrive at the attention of stakeholders and the public (Gottschalk & Benson, 2020), the Fifa scandal is a typical example (Hundt & Horsch, 2019; Naheem, 2018). Persistent public accusations of bribery and corruption forced top officials at Fifa to initiate an investigation. Initially, the Swiss Joseph Blatter, president of Fifa since 1998, denied any wrongdoing. However, Blatter had to resign in 2015 when the Swiss attorney general announced criminal proceedings against him regarding criminal mismanagement and misappropriation. Swiss police arrested seven Fifa officials at hotel Baur au Lac in Zürich. FBI in the United States arrested other Fifa officials and charged them with corruption (Viswanatha et al., 2015).

The 350-page report of investigation mentions Fifa president Joseph 'Sepp' Blatter 81 times. The corporate investigators conclude that Blatter was responsible for mismanagement, but not fraud (Garcia, 2014: 336):

> President Blatter's responsibility for the myriad issues that developed over the course of the bidding process or were uncovered by this inquiry merits consideration. As a preliminary matter, it must be made clear that evidence in the record does not establish a prima facie case that President Blatter violated the FCE. The one concrete allegation against the President, concerning an account purportedly held in his name at a U.S. bank, was demonstrably false. As head of FIFA, however, President Blatter bears some responsibility for a flawed process that engendered deep public skepticism, and for presiding over an Executive Committee whose culture of entitlement contributed to many of the issues this Report identifies.

Fraud examiners argue that Blatter must take responsibility for the failures that occurred on his watch. He made himself accessible on a selective basis, giving the impression that individuals such as Peter Hargitay were insiders afforded preferential treatment, including freedom to speak to high-ranking Fifa officials about inappropriate topics, such as quality of competing bids, in a manner that was not tolerated by others. Hargitay was in the business of lobbying and campaign strategies. He worked as a consultant to several of the nine bidding teams composed of eleven different countries.

Private Policing Evaluation

A review of the investigation report by Garcia (2014) requires evaluation criteria according to the maturity model for private policing in fraud examinations as illustrated in Fig. 11.6. We place the Fifa investigation by Garcia (2014) at level 3 of disclosure in the figure. The desire by powerful Fifa people to keep the investigation report secret is evidence of disclosure of misconduct and potential crime that Fifa executives did not

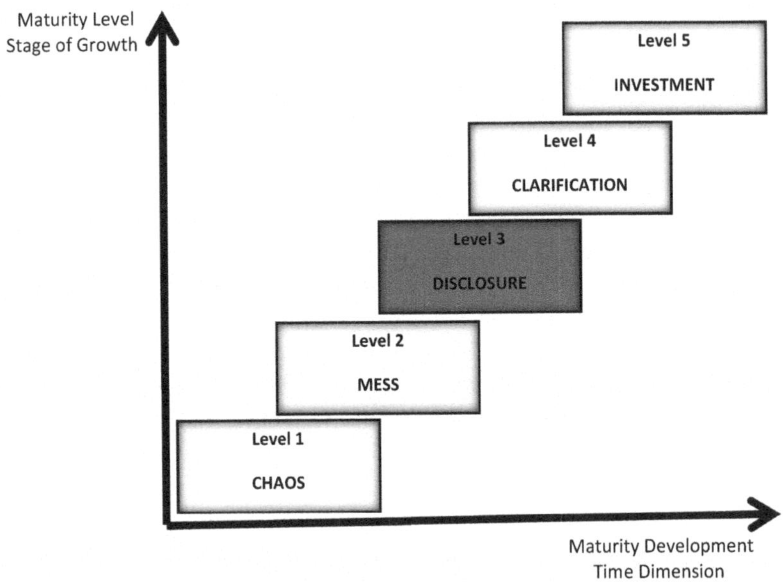

Fig. 11.6 Maturity model for internal private investigations applied to Garcia

want to leak out to governments, the media, and the public in general. The Garcia (2014) report deserves no better evaluation, as there is no real clarification of roles and responsibilities, and there seems to be no learning or other aspects of investment from the report.

Fraud examiners did not succeed in achieving cooperation with important informants. Fraud examiners thus failed, probably because they were confrontational rather than cooperative in their attempts to talk to informants. As mentioned above, there might have been acceptable reasons why executive committee members such as Beckenbauer and Llona did not cooperate with the investigators. If Garcia and the other examiners had approached suspects and witnesses in a more professional manner, they might have succeeded in obtaining relevant and complete statements. Based on their failure to interview informants, the Garcia (2014) investigation has similarities with criteria at level two in the maturity model in Fig. 11.7, as the process of private policing resembled a mess. However, since they succeeded in disclosing new information, the investigation remains at level 3 in the model.

Oceanteam Services by Sands

The background for this private policing of suspected economic crime in terms of an internal investigation by fraud examiners is slightly different from most others, where the organizations themselves initiated investigations. In 2016, one of the minority shareholders in Oceanteam requested an investigation in accordance with Chapter 5 of the Norwegian public limited liability company act. Bergen district court ruled in favor of the demand from the minority shareholder. Oceanteam appealed the ruling. However, in 2018, Gulating court of appeal ruled in accordance with the Bergen district court ruling. Subsequently, lawyer and partner Nicolai Skridshol at the law firm Sands received a request from the court to carry out the investigation. The courts stated the mandate for the assignment given to Sands (2019a, 2019b) in their rulings.

Oceanteam (OT) provided support for offshore contractors all over the world through its fleet of large offshore vessels and its expertise in marine equipment, cable logistics, and design engineering. Chief

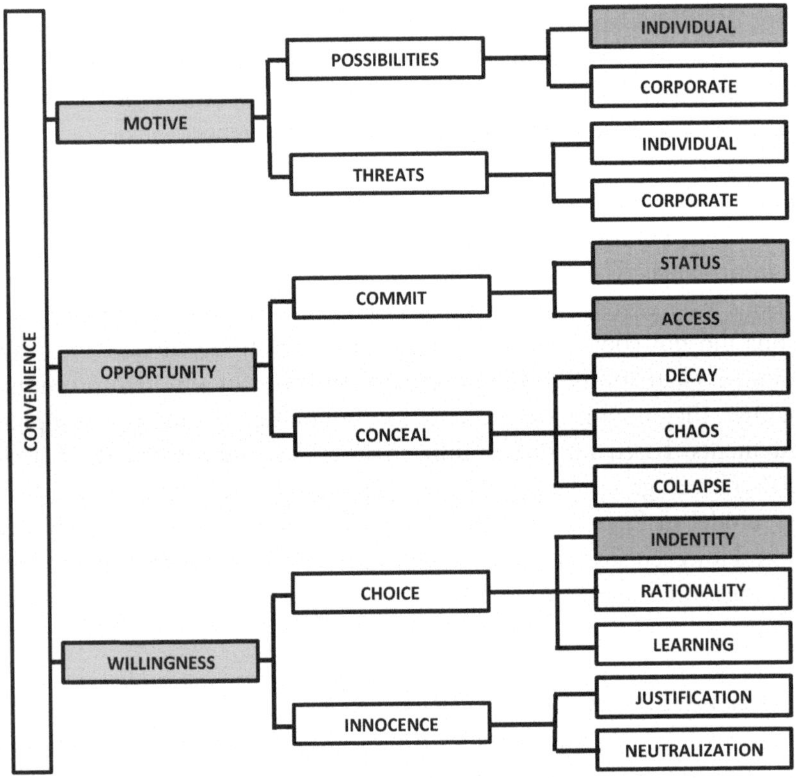

Fig. 11.7 Convenience themes in the case of Oceanteam

executive Haico Halbesma, chief executive, Hessel Halbesma, chairperson, as well as three board members faced accusations of fraud against Oceanteam.

White-Collar Convenience

Assuming the allegations are correct, we briefly study the alleged white-collar crime by application of convenience theory. The chief executive and board members were obviously greedy. Greed implies that offenders are never satisfied, as they always want more profit money (Goldstraw-White, 2012).

The chief executive and board members had status and access (Benson & Simpson, 2018; Cohen & Felson, 1979). Furthermore, it seems that the two Halbesma suffered from narcissistic identification with the organization. Narcissistic identification is a special type of narcissism, where the offender sees little or no difference between self and the corporation. The company money is personal money that can be spent whatever way the narcissist prefers (Galvin et al., 2015). While grandiosity and admiration belong to the motivational dimension of convenience theory, empathy deficit belongs to the willingness dimension of convenience theory where the offender possesses a sense of entitlement (Nichol, 2019). The offender shows unreasonable expectations to receive and obtain preferential treatments (Zvi & Elaad, 2018). Figure 11.7 illustrates narcissistic identification along the axis willingness-choice-identity.

Fraud Examination Outcome

Sands (2019a) found that in general, fraud examiners observed an extensive lack of control of invoices and timesheets from executives and board members. Relevant functions had not controlled board members' timesheets, and the same applied to the CEO. For example, in 2015, Hessel Halbesma invoiced 3.600 additional hours of a total of EUR 1,080,000. Travel bills and other remunerations were extremely high. For example, Hessel Halbesma had travel expenses of EUR 333.000 in 2013 (Sands, 2019a: 9):

> In our review of Hessel Halbesma's travel bills, several expenses classified as company expenses, appear highly questionable with regards to being associated with OT's daily operations, for example purchases of clothing, numerous toll road passings, travels to places where OT has no operations (e.g. Verbier, San Remo), helicopter rental expenses in Monaco and dining expenses with no information about participants, the purpose of the meal or the relation to OT's operations.

Fraud examiners drew similar conclusions regarding Haico Halbesma's travel bills, where several expenses appeared unrelated to Oceanteam's business. Art purchases, sponsorship of a Mexican artist, boat purchase in Monaco, and a personal legal dispute are some of the additional examples presented in the report by Sands (2019a).

Fraud examiners conclude in their report on related party transactions such as fees, salaries, and incentives, travels and remunerations. Other relevant findings and transactions include art purchases (Sands, 2019a: 176):

> We have been informed that Halbesmas allegedly bought art paid by OT (Oceanteam) for private use. However, OT states that they possessed and are still in possession of the art, and that the purchases are registered in B.V. (..) It therefore remains unclear whether OT is in possession of all art specified in the financial account.

It seems surprising that fraud examiners were unable to conclude on this issue. Similarly, fraud examiners failed in concluding whether boat purchases in Monaco and sponsorship of Mexican artist were corporate expenditures or private expenditures paid by the corporation. Yet another example of lacking conclusion is Halbesmas' legal dispute with another party, which Oceanteam paid for (Sands, 2019a: 180):

> Through interviews, we have received various information about what and who this dispute concerns. Both Haico Halbesma and OT's corporate counsel describe the legal matter and the parties involved imprecisely in their interviews and not correctly compared with the legal documents.

The final issue in the factual report is the nomination and remuneration committee. Minority shareholders Storm Capital and Otterley Group had requested such a committee to be established. Over a long period, the shareholders followed up on the matter, but such a committee was never established. Again, fraud examiners fail to conclude on the matter.

Private Policing Evaluation

It seems that Sands' fraud examiners have done a thorough investigation by reconstructing past events and sequences of events related to privileged individuals' abuse of positions at Oceanteam for personal gain. The private policing by Sands (2019a, 2019b) thus deserves an allocation at least at level four in the maturity model in Fig. 11.8. Since there was continued dispute after release of the report, the investment level is not obvious. In fact, after release of the report to Bergen district court, Halbesma and others attempted to prevent shareholders to access the report. While the report was completed and submitted to the court on November 4, 2019, the secrecy went on until April 2020, when a shareholders meeting needed the report. Oceanteam then decided to make the report public on their Web site, commenting that they were not responsible for the contents.

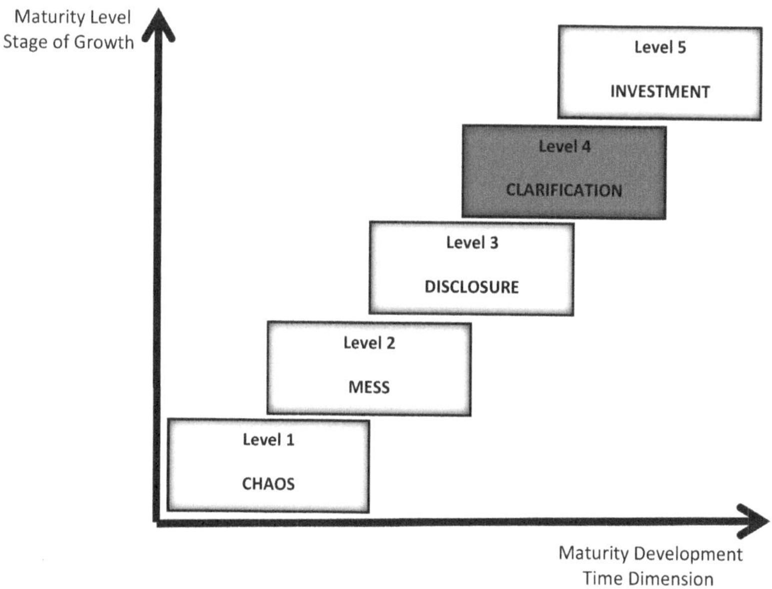

Fig. 11.8 Maturity model for internal private investigations applied to Sands

A letter sent to all shareholders of Oceanteam on March 30, 2020, illustrates the dispute, dissatisfaction, and reluctance to distribute the private policing report, as the letter emphasizes the lack of trust that major shareholders and the company places in the report by Skridshol's team:

> Following a joint petition from the company and the investigation requester, on 4 October 2019, the Bergen district court made a decision to cease the ongoing investigation. In the decision it was further resolved that the investigation report, as far as the report had been written at the time the decision was legally binding, was to be submitted to the Bergen district court.
>
> In accordance with the Public Limited Companies Act section 5-28 second paragraph the Bergen district court convenes the EGM (extraordinary general meeting) to deal with the investigation report (…)
>
> The investigation report has been prepared solely by the investigator, Mr. Nicolai Skridshol, and his investigation team, and the company is not liable for the content of the report. The report was not presented to Oceanteam for review and clarification before being submitted to the court, and the company has not taken any position with respect to the correctness of the factual or legal circumstances described in the report. Oceanteam is in the process of evaluating the report and the company's legal position.

An interesting approach applied by the fraud examiners was to distinguish in their report between a factual review (Sands, 2019a) and a legal review (2019b). The legal review was required in the mandate (Sands, 2019b: 2):

> Legal assessment of the facts, including assessment of whether the mapped transactions violate the provisions of the Public Limited Liability Companies Act on related transactions.

Fraud examiner Skridshol and his team focused on three legal topics: abuse of position, invalidity of agreements due to lack of shareholder approval, and unlawful distributions on the part of Hessel and Haico Halbesma. They identified a number of agreements, transactions, and matters that in their view constitute likely unlawful distributions. Attempts to charge for services not delivered, as well as any other invoices for services not delivered or use of company funds to pay private expenses, can potentially constitute violations of provisions in the penal code.

Moldova Banks by Kroll

This case study is concerned with an internal investigation by fraud examiners in Moldova banks that the National Bank of Moldova initiated. On January 28, 2015, the National Bank of Moldova (NBM) engaged forensics firm Kroll (2016) to conduct a scoping phase of investigation into certain transactions involving Banca de Economii, Banca Sociala, and BC Unibank in Moldova. The Republic of Moldova is a landlocked country in Eastern Europa, bordered by Romania and Ukraine. Kroll is a UK-based firm that conducts investigations.

The New York Times reported in October 2015 that Moldova "was rocked this year by the discovery that $1 billion had fraudulently siphoned from Moldova's banking system over a period of years, a huge amount for an impoverished country whose entire economic output is only about $8 billion a year" (Nechepurenko, 2015). Financial Times reported in January 2016, that, "until the fraud came to light last year, Moldova was considered a rare success in the EU's campaign to build bridges to post-Soviet states."

In 2017, Kroll (2017) was again working for the National Bank of Moldova (NBM) to evidence what caused the three Moldovan banks to collapse, to understand who perpetrated and benefitted from the fraud, and to support criminal and/or civil proceedings to recover the misappropriated funds. One billion dollars had disappeared from the banks in a coordinate effort. Ilan Shor, a 28-year-old Moldovan business executive and the mayor of Orhei organized the swindle. Shor was chairperson of the board of Banca de Economii.

The Times of Israel reported in 2019 that Ilan Shor was sentenced to 7.5 years for fraud and money laundering (Iordachescu & Rodina, 2019; Liphshiz, 2019). Kroll's (2016) scoping phase report found evidence that suggested that Ilan Shor, and companies and individuals affiliated with him (the "Shor Group") played an integral role in coordinating the illegal financial transactions. Kroll (2017) developed further evidence of Shor's involvement leading to his conviction.

The Guardian reported in 2015 that Grant Thornton, the UK-based accountancy firm with local offices in many countries, was the auditor for the three Moldovan banks through which the money was embezzled and spirited out of the country in complex financial transactions. Investigators accused auditing firm Grant Thornton of negligence and incompetence after $1 billion disappeared out of Moldova, a sum equivalent to 15% of the country's gross national product (Rosca, 2015).

White-Collar Convenience

Assuming the allegations are correct, we briefly study the alleged white-collar crime by application of convenience theory. The motive was both individual as well as corporate gain based on possibilities to commit bank fraud, as illustrated in Fig. 11.9. Probably, involved individuals wanted to climb the hierarchy of needs for status and success (Maslow, 1943) and realize the American dream of prosperity (Schoepfer & Piquero, 2006). In addition, they wanted their businesses to make as much profit as possible (Naylor, 2003) and reach business objectives that justify means (Jonnergård et al., 2010).

In the opportunity dimension of convenience theory, Moldovan banks obviously lacked guardianship, control mechanisms, and professional audit functions. There was decay by institutional deterioration (Rodriguez et al., 2005), inability to control because of social disorganization (Hoffmann, 2002), and misrepresentation in accounting (Qiu & Slezak, 2019)—in addition to the status of the offenders in the elite (Pontell et al., 2014).

Elite members in Moldova involved in the bank fraud did it on purpose as a rational choice where perceptions of benefits exceed costs (Pratt & Cullen, 2005). There was probably a behavioral reinforcement of deviance over time (Benartzi et al., 2017), and offenders learned from each other by differential association (Sutherland, 1983). A collectivist

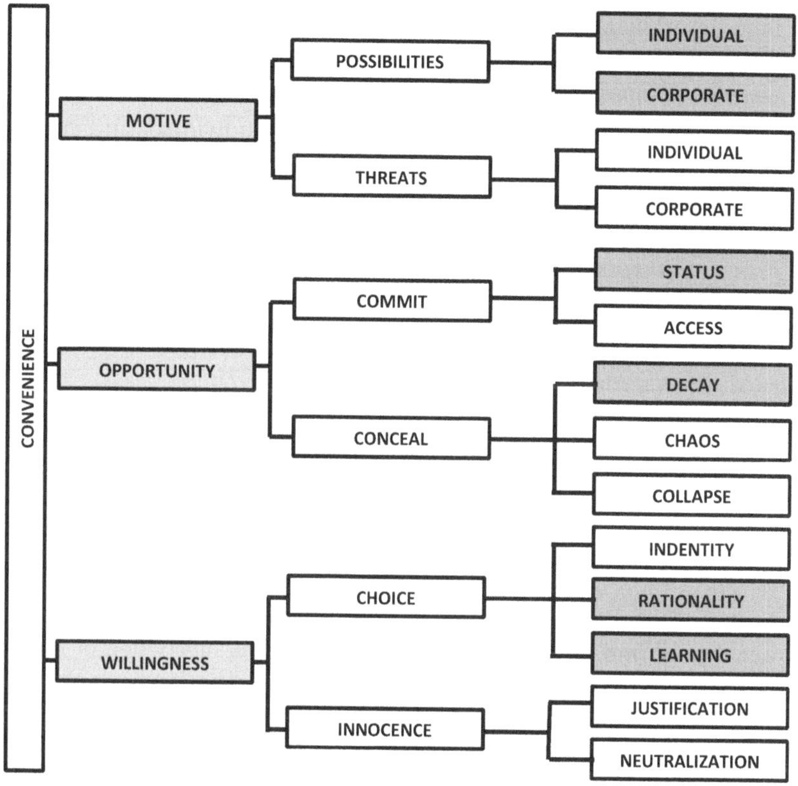

Fig. 11.9 Convenience themes in the case of Moldovan banks

value orientation emerged among offenders (Bussmann et al., 2018), where they perceived no deterrence effect (Comey, 2009).

Fraud Examination Outcome

The mandate for the investigation was to reconstruct events and sequences of events to provide insights into financial irregularities, the main suspect parties, and the roles and responsibilities, In particular, investigators should identify events leading to the collapse of three banks and identify beneficiary of funds that had disappeared from the three banks. The main source of information was NBM.

Fraud examiners from Kroll reviewed transactions from 2012 to 2014 in the three Moldovan banks. The banks were consecutively subject to significant shareholder changes, which had the effect of transferring ownership to a series of apparently unconnected individuals and entities. Each bank entered into a series of transactions resulting in such a significant deterioration in each of their balance sheets that they no longer were viable as going concerns. It was bankruptcy in 2014 in all three banks that triggered the fraud examination.

Fraud examiners established that important bank files were missing. They did not believe the story told to them that someone had stolen and burned down a car with bank files (Kroll, 2016: 67):

> The order was communicated to relevant department heads and 12 sacks of files were transferred to Birca's office. Iurie Buruiana, head of BEM's department for management, collection, and technical protection, was tasked with procuring a vehicle to transport the files to the archive. A vehicle was provided at 7 pm on 26 November by Klassica Force SRL, a security which BEM staff has stated is controlled by Ilan Shor. The sacks were loaded into the vehicle by the driver under the supervision of Iurie Buriana. As it was late, the driver stated that the documents would be transported to the archive the following day, on the 27th, and the vehicle would be held in a secure garage overnight. The following day, Klassica Force SRL informed BEM that the vehicle had been stolen and found burned out, a fact confirmed by a Moldovan police report issued on 1 December 2014.

Fraud examiners suggested the following events leading to the collapse of the banks. First, between January and November 2014, exposure of companies deemed directly and indirectly linked to Ilan Shor increased by MDL three billions to almost MDL eight billions. The fact that almost all of this exposure was concentrated in the three banks cannot be a coincidence. Funding of this explosive loan activity was largely through interbank deposits between the banks.

Second, transactions within each of the three banks had an appearance of lending to related parties to each of them, where the explosive lending activity developed its own dynamics.

Third, such a significant level of interrelated lending with three, apparently independent banks would have required a significant level of co-ordination. Significant connectivity between the shareholders of each of the banks as well as the number of common management changes between the banks could explain the co-ordination.

Fourth, such extended and interrelated funding and loan activity within and between the three banks culminated in a series of events in November 2014, which indicate that this was no longer sustainable. The majority of lending seemed like a routinized activity, facilitated by interbank deposits from Moldovan and Russian banks. Analysis of loan transactions reflects a clear pattern of lending within each bank, finding repayment by loans issued to entities in another bank, via the passing of funds through Russian banks, Latvian bank accounts, and UK limited partnerships.

Fifth, a series of highly unusual events took place on November 28, 2014, at the same time as new loans flowed out of banks, which were further indicative of efforts to extract the maximum value of funds possible from the banks before they collapsed.

Finally, as quoted above, Moldovan police destroyed evidence and reported the loss on December 1.

Private Policing Evaluation

In their second report, Kroll (2017: 20) admits that they failed in interviewing the main suspect Ilan Shor, and they failed in obtaining a verifiable statement from him, which is an obvious shortcoming of the investigation:

> Kroll was made aware of the existence of a statement that has been uploaded onto the internet, which purports to be Mr. Shor's statement to the Public Prosecutor, providing an explanation for some of the events that took place within the Three Moldovan Banks. Kroll has not been provided with an official copy of Mr. Shor's statement and has not had the opportunity to meet with Mr. Shor. As such we are not at this stage in a position to confirm the authenticity of the document and do not comment on its content within this report.

The quoted statement seems arrogant and ignoring criminal justice, where everyone is innocent until proven guilty. Furthermore, fraud examiners should have listened to and respected statements from a suspect, since everyone is innocent until proven guilty, and not ignored as seems to be the case by fraud examiners from Kroll. Therefore, we consider their two investigations far from perfect. A professional crime investigation is about not only what you find out, but also how you did it, and how you treated suspects and witnesses in the process.

Fraud examiners relied too heavily on information from NBM, especially in the first investigation (Kroll, 2016: 7):

> In conducting this preliminary scoping phase of investigation, Kroll has relied upon information and analysis conducted by NBM, in order to conduct an assessment of the basis of their findings to date, and concerns raised.

Nevertheless, we place the Kroll (2016, 2017) reports at the clarification level 4 in the maturity model in Fig. 11.10.

In conclusion, when there is suspicion of economic crime by white-collar offenders in organizations, the organizations tend to hire fraud examiners to conduct private policing by internal investigations in the organizations. The affected organizations become clients of law firms and auditing firms who reconstruct past events and sequences of events to determine the extent of misconduct and crime. This chapter has presented a few relevant case studies to illustrate the extent of convenience in white-collar crime and the level of maturity in fraud examinations. Overall, most internal investigators have a long way to go before their work deserves the classification of an investment for the clients. Failing interviews, lack of relevant knowledge, and conclusions without reliable evidence are some of the shortcomings identified in the case studies. As pointed out in previous research (e.g., Brooks & Button, 2011; Button & Gee, 2013; Button et al., 2007a, 2007b; Gottschalk, 2020; Schneider, 2006; Williams, 2005a, 2005b), many fraud examiners have a long way to go before they can deserve the label professional investigators or examiners. Future research may help professionalize the business of private policing of economic crime.

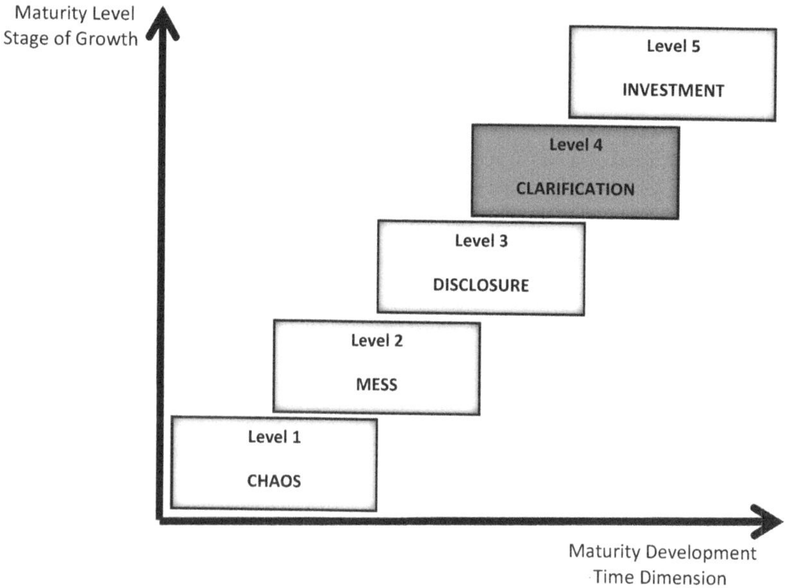

Fig. 11.10 Maturity model for internal private investigations applied to Kroll

Bibliography

Agnew, R. (2014). Social concern and crime: Moving beyond the assumption of simple self-interest. *Criminology, 52*(1), 1–32.

Bell, E., Bryman, A., & Harley, B. (2018). *Business research methods* (2nd ed.). Oxford University Press.

Benartzi, S., Beshears, J., Milkman, K. L., Sunstein, C. R., Thaler, R. H., Shankar, M., et al. (2017). Should governments invest more in nudging? *Psychological Science, 28*(8), 1041–1055.

Benson, M. L., & Simpson, S. S. (2018). *White-collar crime: An opportunity perspective* (3rd ed.). Routledge.

Braaten, C. N., & Vaughn, M. S. (2019). Convenience theory of cryptocurrency crime: A content analysis of U.S. federal court decisions. *Deviant Behavior*, published online https://doi.org/10.1080/01639625.2019.1706706.

Brooks, G., & Button, M. (2011). The police and fraud investigation and the case for a nationalized solution in the United Kingdom. *The Police Journal, 84,* 305–319.

Bussmann, K. D., Niemeczek, A., & Vockrodt, M. (2018). Company culture and prevention of corruption in Germany, China and Russia. *European Journal of Criminology, 15*(3), 255–277.

Button, M. (2019). The "new" private security industry, the private policing of cyberspace and the regulatory questions. *Journal of Contemporary Criminal Justice,* published online. https://doi.org/10.1177/1043986219890194.

Button, M., & Gee, J. (2013). *Countering fraud for competitive advantage—The professional approach to reducing the last great hidden cost*. Wiley.

Button, M., Frimpong, K., Smith, G., & Johnston, L. (2007a). Professionalizing counter fraud specialists in the UK: Assessing progress and recommendations for reform. *Crime Prevention and Community Safety, 9,* 92–101.

Button, M., Johnston, L., Frimpong, K., & Smith, G. (2007b). New directions in policing fraud: The emergence of the counter fraud specialists in the United Kingdom. *International Journal of the Sociology of Law, 35*, 192–208.

Chan, F., & Gibbs, C. (2020). Integrated theories of white-collar and corporate crime. In M. L. Rorie (Ed.), *The handbook of white-collar crime*. Wiley, chapter 13, pp. 191–208.

Chatterjee, A., & Pollock, T. G. (2017). Master of puppets: How narcissistic CEOs construct their professional worlds. *Academy of Management Review, 42*(4), 703–725.

Clifford Chance. (2020). *Report of investigation on Swedbank*, March 23, law firm Clifford Chance, Washington, DC.

Cohen, L. E., & Felson, M. (1979). Social change and crime rate trends: A routine activity approach. *American Sociological Review, 44,* 588–608.

Comey, J. B. (2009). Go directly to prison: White collar sentencing after the Sarbanes-Oxley act. *Harvard Law Review, 122,* 1728–1749.

Eabrasu, M. (2020). Cheating in business: A metaethical perspective. *Journal of Business Ethics, 162,* 519–532.

Galvin, B. M., Lange, D., & Ashforth, B. E. (2015). Narcissistic organizational identification: Seeing oneself as central to the organization's identity. *Academy of Management Review, 40*(2), 163–181.

Gamache, D. L., & McNamara, G. (2019). Responding to bad press: How CEO temporal focus influences the sensitivity to negative media coverage of acquisitions. *Academy of Management Journal, 62*(3), 918–943.

Garcia. (2014). *Report on the Inquiry into the 2018/2019 Fifa World Cup Bidding Process*, Investigatory Chamber, FIFA Ethics Committee, Zürich, Switzerland.

Goldstraw-White, J. (2012). *White-collar crime: Accounts of offending behavior*. Palgrave Macmillan.

Gottschalk, P. (2019). *Convenience triangle in white-collar crime–Case studies of fraud examinations*. Edward Elgar Publishing.

Gottschalk, P. (2020). Private policing of white-collar crime: Case studies of internal investigations by fraud examiners. *Police Practice and Research*, published online. https://doi.org/10.1080/15614263.2020.1789461.

Gottschalk, P., & Benson, M. L. (2020). The evolution of corporate accounts of scandals from exposure to investigation. *British Journal of Criminology, 60*, 949–969.

Gottschalk, P., & Tcherni-Buzzeo, M. (2017). Reasons for gaps in crime reporting: The case of white-collar criminals investigated by private fraud examiners in Norway. *Deviant Behavior, 38*(3), 267–281.

Hansen, L. L. (2020). Review of the book "Convenience triangle in white-collar crime: Case studies of fraud examinations". *ChoiceConnect, 57*(5). Association of College and Research Libraries.

Hoffmann, J. P. (2002). A contextual analysis of differential association, social control, and strain theories of delinquency. *Social Forces, 81*(3), 753–785.

Hundt, S., & Horsch, A. (2019). Sponsorship of the Fifa world cup, shareholder wealth, and the impact of corruption. *Applied Economics, 51*(23), 2468–2491.

Iannacci, F., Seepma, A. P., Blok, C., & Resca, A. (2019). Reappraising maturity models in e-government research: The trajectory-turning point theory. *Journal of Strategic Information Systems, 28*, 310–329.

Iordachescu, I., & Rodina, M. (2019). Israeli-born convicted fraudster on track for seat in Moldovan parliament. *The Times of Israel*. www.timesofisrael.com, published February 20.

Jonnergård, K., Stafsudd, A., & Elg, U. (2010). Performance evaluations as gender barriers in professional organizations: A study of auditing firms. *Gender, Work and Organization, 17*(6), 721–747.

King, M. (2020a). What makes a successful corporate investigator—An exploration of private investigators attributes. *Journal of Financial Crime*, published online https://doi.org/10.1108/JFC-02-2020-0019.

King, M. (2020b). Out of obscurity: The contemporary private investigator in Australia. *International Journal of Police Science and Management*, published online. https://doi.org/10.1177/1461355720931887.

Kireenko, A. P., Nevzorova, E. N., & Fedotov, D. Y. (2019). Sector-specific characteristics of tax crime in Russia. *Journal of Tax Reform, 5*(3), 249–264.

Krippendorff, K. (1980). *Content analysis: An introduction to its methodology*. Sage.

Kroll. (2016). *Project Tenor—Scoping Phase, Final Report prepared for The National Bank of Moldova.* April 2, investigation firm Kroll, Nexus Place, London, UK.

Kroll (2017). *Project Tenor II—Summary Report, Report prepared for The National Bank of Moldova.* December 20, investigation firm Kroll, Nexus Place, London, UK.

Laughland, O. (2017). Fifa official took bribes to back Qatar's 2022 world cup bid, court hears. *The Guardian*. www.theguardian.com, published November 15.

Leasure, P., & Zhang, G. (2018). "That's how they taught us to do it": Learned deviance and inadequate deterrents in retail banking. *Deviant Behavior, 39*(5), 603–616.

Liphshiz, C. (2019). Moldova Jews bear anti-semitic brunt as corrupt lawmaker flees, maybe to Israel. *The Times of Israel*. www.thetimesofisrael.com, published August 30.

Makortoff, K. (2019). Swedbank chief sacked amid money laundering scandal. *The Guardian*. www.theguardian.com, published March 28.

Maslow, A. H. (1943). A theory of human motivation. *Psychological Review, 50*(4), 370–396.

McClelland, P. L., Liang, X., & Barker, V. L. (2010). CEO commitment to the status quo: Replication and extension using content analysis. *Journal of Management, 36*(5), 1251–1277.

Milne, R. (2020). Swedbank failings on E37bn of transactions revealed in report. *Financial Times*. www.ft.com, published March 23.

Naheem, M. A. (2018). Fifa—Highlighting the links between global banking and international money laundering. *Journal of Money Laundering Control, 21*(4), 498–512.

Naylor, R. T. (2003). Towards a general theory of profit-driven crimes. *British Journal of Criminology, 43*, 81–101.

Nechepurenko, I. (2015). Moldova parliament dismisses government amid bank scandal. *The New York Times*. www.nytimes.com, published October 29.

Nichol, J. E. (2019). The effects of contract framing on misconduct and entitlement. *The Accounting Review, 94*(3), 329–344.

Otu, S. E., & Okon, O. N. (2019). Participation in fraud/cheat in the buying and selling of meats without legal metrology: A theoretical and empirical investigation. *Deviant Behavior, 40*(2), 205–224.

Patrucco, A. S., Luzzini, D., & Ronchi, S. (2017). Research perspectives on public procurement: Content analysis of 14 years of publications in the Journal of Public Procurement. *Journal of Public Procurement, 16*(2), 229–269.

Pontell, H. N., Black, W. K., & Geis, G. (2014). Too big to fail, too powerful to jail? On the absence of criminal prosecutions after the 2008 financial meltdown. *Crime, Law and Social Change, 61*(1), 1–13.

Pratt, T. C., & Cullen, F. T. (2005). Assessing macro-level predictors and theories of crime: A meta-analysis. *Crime and Justice, 32*, 373–450.

Qiu, B., & Slezak, S. L. (2019). The equilibrium relationships between performance-based pay, performance, and the commission and detection of fraudulent misreporting. *The Accounting Review, 94*(2), 325–356.

Reese, B., & McDougal, M. K. (2018). Gender, status, and tax offenses. *Deviant Behavior, 39*(12), 1647–1657.

Rodriguez, P., Uhlenbruck, K., & Eden, L. (2005). Government corruption and the entry strategies of multinationals. *Academy of Management Review, 30*(2), 383–396.

Rosca, M. (2015). Vanishing act: How global auditor failed to spot theft of 15% of Moldova's wealth. *The Guardian*. www.theguardian.com, published July 1.

Röglinger, M., Pöppelbuss, J., & Becker, J. (2012). Maturity model in business process management. *Business Process Management Journal, 18*(2), 328–346.

Sands. (2019a). *Factual report: Oceanteam ASA investigation of related party transactions*. November 4, law firm Sands, Oslo, Norway.

Sands. (2019b). *Report—Legal review: Oceanteam ASA investigation of related party transactions*, November 4, law firm Sands, Oslo, Norway.

Saunders, M., Lewis, P., & Thornhill, A. (2007). *Research methods for business students* (5th ed.). Pearson Education.

Schnatterly, K., Gangloff, K. A., & Tuschke, A. (2018). CEO wrongdoing: A review of pressure, opportunity, and rationalization. *Journal of Management, 44*(6), 2405–2432.

Schneider, S. (2006). Privatizing economic crime enforcement: Exploring the role of private sector investigative agencies in combating money laundering. *Policing & Society, 16*(3), 285–312.

Schoepfer, A., & Piquero, N. L. (2006). Exploring white-collar crime and the American dream: A partial test of institutional anomie theory. *Journal of Criminal Justice, 34*(3), 227–235.

Solli-Sæther, H., & Gottschalk, P. (2015). Stages-of-growth in outsourcing, offshoring and backsourcing: Back to the future? *Journal of Computer Information Systems, 55*(2), 88–94.

Sutherland, E. H. (1983). *White collar crime—The uncut version.* Yale University Press.

Sykes, G. M., & Matza, D. (1957). Techniques of neutralization: A theory of delinquency. *American Sociological Review, 22*(6), 664–670.

Vasiu, V. I., & Podgor, E. S. (2019, July). Organizational opportunity and deviant behavior: Convenience in white-collar crime. *Criminal Law and Criminal Justice Books.* Rutgers, the State University of New Jersey. www.clcjbooks.rutgers.edu.

Viswanatha, A., Robinson, J., Morse, A., & Matthews, C. M. (2015). Fifa rocked as U.S. charges 14 in corruption investigation. *The Wall Street Journal*. www.wsj.com, published May 27.

Williams, J. W. (2005a). Reflections on the private versus public policing of economic crime. *British Journal of Criminology, 45,* 316–339.

Williams, J. W. (2005b). Governability matters: The private policing of economic crime and the challenge of democratic governance. *Policing & Society, 15*(2), 187–211.

Wood, J. D. (2020). Private policing and public health: A neglected relationship. *Journal of Contemporary Criminal Justice, 36*(1), 19–38.

Zhu, D. H., & Chen, G. (2015). CEO narcissism and the impact of prior board experience on corporate strategy. *Administrative Science Quarterly, 60*(1), 31–65.

Zvi, L., & Elaad, E. (2018). Correlates of narcissism, self-reported lies, and self-assessed abilities to tell and detect lies, tell truths, and believe others. *Journal of Investigative Psychology and Offender Profiling, 15,* 271–286.

12

CEO Crime Case Study

This chapter presents a case study of traditional white-collar crime by a self-proclaimed environmentalist in Sweden. The case study illustrates how the structural model of convenience theory can be applied to a specific white-collar offender to identify relevant convenience themes that can explain the occurrence of crime. The case study is based on archival analysis of Swedish court documents, fraud investigation reports, an autobiography by the former CEO, and media reports. His motive seems dominated by individual and corporate possibilities as well as corporate threats. His opportunity seems dominated by status, access, and chaos. His willingness seems dominated by identity and justification. A white-collar offender is a person in the upper echelon of society, such as a corporate chief executive officer (CEO), who abuses the professional position to commit and conceal financial wrongdoing (Piquero, 2012, 2018; Schoultz & Flyghed, 2020a, 2020b; Stadler et al., 2013; Sutherland, 1939, 1983). An example of misconduct is accounting manipulation, where an offender misrepresents figures and distorts presentations for income, costs, or assets (Qiu & Slezak, 2019). Executives might enter income too early into financial statements. For example, CEO Whittaker at Fuji Xerox in New Zealand launched an aggressive sales practice where

P. Gottschalk, *Detecting and Investigating Environmental Crime*,
https://doi.org/10.1007/978-3-030-74184-6_12

clients had to sign new contracts before old contracts expired (Deloitte, 2017). The income from new contracts was immediately registered in financial statements (Hamish, 2016; Henderson, 2016).

The emerging theory of convenience suggests that a strong motive, an attractive opportunity, and a high-level willingness for deviant behavior increase the tendency of misconduct, wrongdoing, and crime by persons in the privileged elite (Braaten & Vaughn, 2019; Dearden & Gottschalk, 2020). Convenience is a concept that was theoretically mainly associated with efficiency in time savings. Today, convenience is associated with a number of other characteristics, such as reduced effort and reduced pain. Convenience is associated with terms such as fast, easy, and safe. Convenience says something about attractiveness and accessibility. A convenient individual is not necessarily neither bad nor lazy. On the contrary, the person can be seen as smart and rational (Sundström & Radon, 2015).

This chapter applies convenience theory to a convicted CEO in Sweden. The theory suggests a number of potential convenience themes for financial motive, organizational opportunity, and personal willingness. Some of the themes are relevant to the case study CEO. The purpose of this chapter is to illustrate how convenience themes in the emerging theory can be selected in a specific case study of a white-collar offender.

Literature Review

As described by Chan and Gibbs (2020), the theory of convenience is an integrated deductive approach to explain white-collar crime. Vasiu and Podgor (2019) argue that the theory of convenience provides a new landscape for evaluating white-collar offences. Hansen (2020) suggests that the theory of convenience is based on a new theoretical paradigm. Braaten and Vaughn (2019) found support for convenience theory when studying cryptocurrency crime, while Dearden and Gottschalk (2020) explain white-collar crime target selection by convenience theory.

Convenience is the state of being able to proceed with something with little effort or difficulty, avoiding pain and strain (Mai & Olsen, 2016).

Convenience is savings in time and effort (Farquhar & Rowley, 2009), as well as avoidance of pain and obstacles (Higgins, 1997). Convenience is a relative concept concerned with efficiency in time and effort as well as reduction in pain and solution to problems (Engdahl, 2015). Convenience is an advantage in favor of a specific action to the detriment of alternative actions. White-collar offenders choose the most convenient path to reach their goals (Wikstrom et al., 2018).

The integrated deductive theory of convenience results from a synthesis of perspectives in three dimensions as illustrated in Figure 12.1:

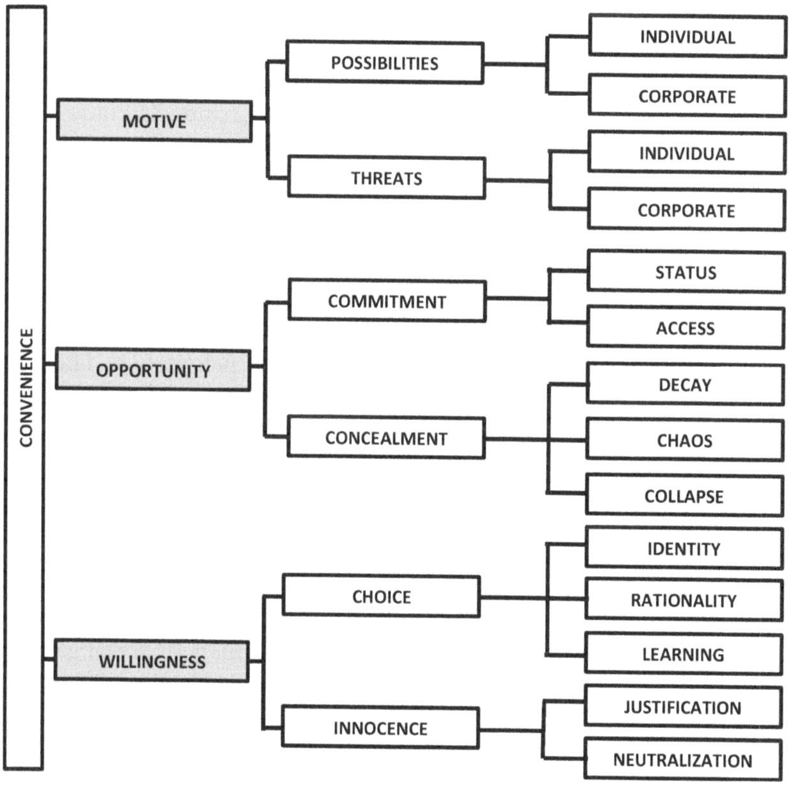

Fig. 12.1 Structural model of convenience theory

- *Convenience in motive*. It is convenient to use illegitimate financial gain to explore possibilities and avoid threats. Climb the hierarchy of needs for status and success (Maslow, 1943), realize the American dream of prosperity (Schoepfer & Piquero, 2006), satisfy the need for acclaim as a narcissist (Chatterjee & Pollock, 2017), restore the perception of equity and equality (Leigh et al., 2010), and satisfy the desire to help others as social concern (Agnew, 2014) are themes integrated in the individual possibilities perspective of convenience theory. Reach business objectives that justify means (Jonnergård et al., 2010), satisfy greed where nothing is ever enough (Goldstraw-White, 2012), enjoy mutual benefits in exchange relationships (Huang & Knight, 2017), and make as much profit as possible as the only goal (Naylor, 2003) are themes integrated in the corporate possibilities perspective. Avoid loss of self-esteem after organizational failure (Crosina & Pratt, 2019), remove strain pain, and uncertainty (Langton & Piquero, 2007), and avoid falling from position in the privileged elite (Piquero, 2012) are themes integrated in the individual threats perspective of convenience theory. Avoid corporate collapse and bankruptcy (Blickle et al., 2006), adapt to profitable market forces (Leonard & Weber, 1970), and join profitable criminal networks and cartels (Goncharov & Peter, 2019) are themes integrated in the corporate threats perspective. The extra profit from financial crime enables the offender to handle desired possibilities and potential threats. It is mainly the convenience of extra profit, rather than the convenience of illegal profit, that is important in the motive dimension of convenience theory. However, under certain circumstances, there might be some extra benefits from illegal extra profit compared to legal extra profit, since illegal funds avoid the attention of external and internal control mechanisms, including compliance functions (Kawasaki, 2020). Illegitimate financial gain can thus find its way into exploring possibilities and avoiding threats that recorded funds cannot.
- *Convenience in opportunity*. There is convenient access to resources to commit and conceal financial crime. Status opportunity derives from executive language that people do not understand (Ferraro et al., 2005), elite members that are too big to fail and too powerful to jail (Pontell et al., 2014), blame game by misleading attribution to

others (Eberly et al., 2011), offender humor distraction from deviant behavior (Yam et al., 2018), and power inequality between the elite and others (Patel & Cooper, 2014). Access opportunity derives from legitimate access to premises and systems (Benson & Simpson, 2018), opportunity creation by entrepreneurship (Ramoglou & Tsang, 2016), specialized access in routine activity (Cohen & Felson, 1979), and legitimate access to strategic resources (Adler & Kwon, 2002). Decay opportunity derives from institutional deterioration based on legitimacy (Rodriguez et al., 2005), inability to control because of social disorganization (Hoffmann, 2002), interference and noise in crime signals (Karim & Siegel, 1998), misrepresentation in accounting (Qiu & Slezak, 2019), and auditors reporting to management rather than the board (Hurley et al., 2019). Chaos opportunity derives from lack of control in principal-agent relationships (Bosse & Phillips, 2016), sense-making of actions difficult for outsiders (Weick, 1995), costs exceed benefits for whistleblowers (Keil et al., 2010), and ethical climate conflict (Victor & Cullen, 1988). Collapse opportunity derives from rule complexity preventing compliance (Lehman et al., 2019), participation in crime networks such as cartels (Nielsen, 2003), and usual way of business in markets with crime forces (Chang et al., 2005). A typical white-collar offender does not go into hiding as many street criminals do. Rather, the offender conceals financial crime among legal transactions to make illegal transactions seem legitimate, or the offender conceals financial crime by removing certain activities from the books. A typical white-collar offender who has convenient legitimate access to commit crime might spend most of the energy on concealing crime in the professional context (Huisman & Erp, 2013; McClanahan & South, 2020).

- *Convenience in behavior*. Offenders can conveniently justify crime and neutralize guilt feelings. In the perspective of identity, themes in convenience theory include the acceptability for the elite from social conflict (Petrocelli et al., 2003), narcissistic identification with the organization (Galvin et al., 2015), professional deviant identity (Obodaru, 2017), narcissist expecting preferential treatment (Zvi & Elaad, 2018), reputation adaption to individual labels (Bernburg et al, 2006), social ties that dwindle with age (Sampson & Laub, 1993), and deviant identity labeling (Mingus & Burchfield, 2012). In the perspective of rationality, themes in convenience theory include perception of benefits exceeding costs (Pratt & Cullen, 2005), behavioral reinforcement of deviance over time (Benartzi et al., 2017), undesirable impulses in self-regulation (Mawritz et al., 2017), no perceived deterrence effect (Comey, 2009), work-related stress that is self-determined (Olafsen et al., 2017), and sensation seeking to experience adventure (Craig & Piquero, 2017). In the perspective of learning, themes include learning from others by differential association (Sutherland, 1939, 1983), action according to authority as obedience (Baird & Zelin, 2009), and collectivist value orientations (Bussmann et al., 2018). In the perspective of justification, themes include the act of wrongdoing as morally justifiable (Schnatterly et al., 2018), upper echelon information selection (Gamache & McNamara, 2019), disappointing work context cause entitlement (Nichol, 2019), negative life events have occurred (Engdahl, 2015), and peer pressure (Gao & Zhang, 2019). In the perspective of neutralization, themes in convenience theory include application of neutralization techniques (Sykes & Matza, 1957), sliding on the slippery slope (Welsh et al., 2014), and lack of self-control (Gottfredson & Hirschi, 1990). When a white-collar offender denies a guilty mind, then the offender applies neutralization techniques. When a white-collar offender makes crime as a choice, it is convenient based on identity (Galvin et al., 2015), rationality (Pratt & Cullen, 2005), and learning from others (Sutherland, 1983).

The convenience triangle for white-collar crime has similarities with the fraud triangle (Cressey, 1972), which suggests three conditions for fraud: (1) incentives and pressures, (2) opportunities, and (3) attitudes and rationalization. However, there are three distinct differences. First, convenience is a relative concept, indicating that offenders have the option of alternative actions to reach their goals that do not represent illegitimate behavior. While the fraud triangle suggests that opportunities will stimulate crime, the convenience triangle suggests that relative opportunities will stimulate crime. It is crime-as-a-choice. There is no reason to commit crime, even if there are many opportunities, as long as alternative decisions may lead to the same result. It is the extent of convenience, and not the extent of opportunity, that determines whether an offense is attractive. A very conveniently oriented decision-maker may resort to illegal activities when legal activities are slightly more stressful. A less conveniently oriented decision-maker may try intensely to solve problems and explore opportunities without violating the law.

Second, it is in the organizational setting where offenders have access to resources so that opportunity arises to commit and conceal crime. While the fraud triangle emphasizes opportunity in general, the convenience triangle concentrates on the privileged position that offenders can abuse to commit and conceal crime. There is trust and lack of control, obedience and fear, which create convenient opportunities. The convenient opportunity derives from legitimate access to resources in a trusted position without guardians, where resources are enablers to carry out activities that are not available to others. Opportunity convenience emerges because of an organizational structure and an organizational culture where members of the elite may feel above the law.

Third, a white-collar offender can influence the organizational opportunity over time. Therefore, opportunity in convenience theory is a dynamic rather than a static condition. By collecting decision rights, by controlling information flows, and by authoritarian leadership styles a potential offender develops an opportunity space that grows over time. Whether intentional or not, the opportunity space changes over time as a reaction to the potential offender's behavior.

This chapter presents an application of convenience theory to accounting manipulation by a white-collar offender. Manipulation and misreporting in accounting is often a convenient way of concealing illegal transactions and presenting a distorted picture of the real financial situation (Qiu & Slezak, 2019). The offender misrepresents figures and distorts presentations of income, costs, and assets with a purpose of creating a false impression of an organization's financial strength, as exemplified by CEO Whittaker at Fuji Xerox in the introduction.

Lack of transparency makes concealment in accounting convenient (Davidson et al., 2019; Goncharov & Peter, 2019). In addition, reporting accounting fraud to public authorities will also harm auditors (Mohliver, 2019: 316):

> As organizations, audit firms are often severely penalized for client malfeasance. Yet the individual auditors working for these firms are susceptible to "motivated blindness" stemming from conflicts of interest that bias their moral judgment toward choices that help their clients.

Mohliver (2019) found that auditor bias toward accepting deviant financial reports increased when there is ambiguity about the appropriateness of a course of action. Financial misreporting that is viewed favorably by the client organization can be recommended by external auditors on the grounds that such reporting is already adopted among companies served by the same auditing firm.

Alon et al. (2019) argue that accounting and auditing functions have undergone a legitimacy crisis in recent years. Auditors are supposed to serve as gatekeepers to protect shareholders and other stakeholders, but deviant corporate management tend to hire and control auditors instead of letting auditors report to the board of directors or the supervisory board (Hurley et al., 2019).

Research Method

This research is concerned with application of convenience theory to a convicted chief executive officer in Sweden. The CEO was convicted of misrepresentation in accounting. The research method is archival study of Swedish court documents (Södertörn, 2015, Svea, 2017), fraud investigation reports (Ernst & Young, 2012; Nergaard, 2013), an autobiography (Benulic, 2018), and media reports (Mo, 2018; Nødset, 2012).

The research technique applied in this empirical study of archival material is content analysis (Bell et al., 2018; Saunders et al., 2007). Content analysis is any methodology or procedure that works to identify characteristics within texts attempting to make valid inferences (Duriau et al., 2007; Krippendorff, 1980; Patrucco et al., 2017). Content analysis assumes that language reflects both how people understand their surroundings and their cognitive processes (Hsieh & Shannon, 2005). Cognition refers to what people think and how they think, and cognitive processes affect the way in which people interpret and make sense of what is around them. Therefore, content analysis makes it possible to identify and determine relevant text in a context (McClelland et al., 2010).

Boris Benulic was the chief executive officer (CEO) at energy company Kraft & Kultur in Sweden, which was a subsidiary of the Norwegian energy corporation Troms Kraft (Nergaard, 2013). It seems that Benulic made the choice of crime based on a perception of benefits exceeding costs as rational, and it seems that behavioral reinforcement of deviance took place over time. His salary increased substantially, as board members believed they observed growth in sales and profits, especially growth in financial contributions to the group and parent company, and the board paid him bonuses for seemingly successful entrepreneurship (Benulic, 2018).

Troms Kraft hired corporate investigators from global auditing firm Ernst & Young (2012) to conduct an internal investigation at Kraft & Kultur. They were surprised to find a series of emails from Benulic instructing accountants to manipulate figures in such ways that the bottom lines in terms of profits would result in predetermined numbers. For example, in an email dated June 5, 2008, Benulic wrote that the electricity income had to be set in such a way that the income from electricity would result in a sum that made the firm complete at 16,1 million Swedish kroner in profits after tax (Ernst & Young, 2012: 10):

> Elintäkterna – ta en summa där som gör at vi landar på 16,1 MSEK i resultat etter skatt (Electricity income – take a sum that makes us landing at 16,1 million Swedish kroner in result after tax).

In 2018, the Supreme Court in Sweden upheld the final verdict of three and a half years in prison for Boris Benulic for accounting fraud (Mo, 2018), which had been determined by the district court of Södertörn (2015) and the court of appeal Svea (2017).

Research Results

Boris Benulic argued in court that the prosecution had made allegations based on incorrect and defective investigations. Among other things, the prosecutor did not take into consideration that Kraft & Kultur had different types of electricity agreements, nor did the prosecutor consider the in-depth balance on the issue of accrued income. He further pointed out that he, as the defendant, suffered an unfair trial. He referred to guidelines that a trial must be impartial and independent. Investigator Annicka Holmgren, who the court appointed as an expert, was not objective according to Benulic, because she had worked for Troms Kraft. The company had paid for her investigation and the expertise she stands for.

Benulic blamed start-up problems that Kraft & Kultur had with regard to difficulties in linking electricity consumption to the correct period during billing, and that there had therefore been a transfer of billing, in some cases over several years. He also pointed out that customers had not reported power consumption on time, which contributed to shifts in power costs. Nor with the help of Enita AS, the company that was supposed to help with structuring and consulting, did Kraft & Kultur obtain a complete and correct annual report on accrued revenues. Benulic thus claimed that he had tried as far as he could to make the accounting items for earned annual income as accurate as possible. He blamed low effectiveness and quality of internal controls for potential errors.

In an interview, Benulic envisioned himself as the CEO of Kraft & Kultur to be a major enterprise with headquarters in New York. This illustrates his desire to grow the company and run a large organization (Nødset 2012: 58):

> The motivation to grow is what usually makes a good CEO. Due to the high competitive pressure, the company had to offer extremely low prices in order to grow and outbid competitors. K&K's competitors could not understand how the company was able to offer such low prices.

In addition to growth and fame, Benulic was also interested in personal financial benefits. When the company seemingly doubled its sales from

2008 and 2010, Benulic's salary almost doubled from nine hundred thousand Swedish kroner to one million and six hundred thousand kroner. While Figure 11.2 illustrates growth by motive-possibilities-corporate, the figure illustrates greed by motive-possibilities-individual.

At the same time, there is also motive-threats-corporate, where the chief executive registered "fake future income that was used to hide losses at K&K" (Nergaard, 2013: 277).

While Benulic justified his actions as illustrated in Figure 11.2, he also took on the sole responsibility for any mistakes in an email to the board on November 12, 2011, two weeks before the board fired him (Ernst & Young, 2012: 10):

> The chief executive writes in the email that nobody else in the company was involved, and that the mistake is his responsibility alone.

Benulic was a well-known person in Sweden because of his career background as a media reporter, investigative journalist, newspaper editor, participant in public debates, and left-wing politician. People perceived him as charismatic and influential with personal skills to be successful. However, he implemented an authoritarian management style when he took on the leadership at Kraft & Kultur (Nødset, 2012).

As illustrated in Figure 11.2, Benulic had high status and convenient access to resources by telling people what to do. He had control over the external auditor Elisabeth Simonssons from auditing firm Grant Thornton (Ernst & Young, 2012).

Benulic had a strong identification with the business and justified his wrongdoings. Nergaard (2013) suggests that the opportunity structure also consisted of weak management competence in the parent company Troms Kraft, incompetent and marginal finance function, minimal and absent control systems, and board members knowing energy but not business.

In the willingness dimension of convenience theory in Figure 12.2, Benulic came to Kraft & Kultur with a tendency to act on both the wrong side and the right side of the law in his previous positions. For example, as a publisher at Ordfront, he regularly exceeded his authorization, was reluctant to inform the board about new contracts, and let the

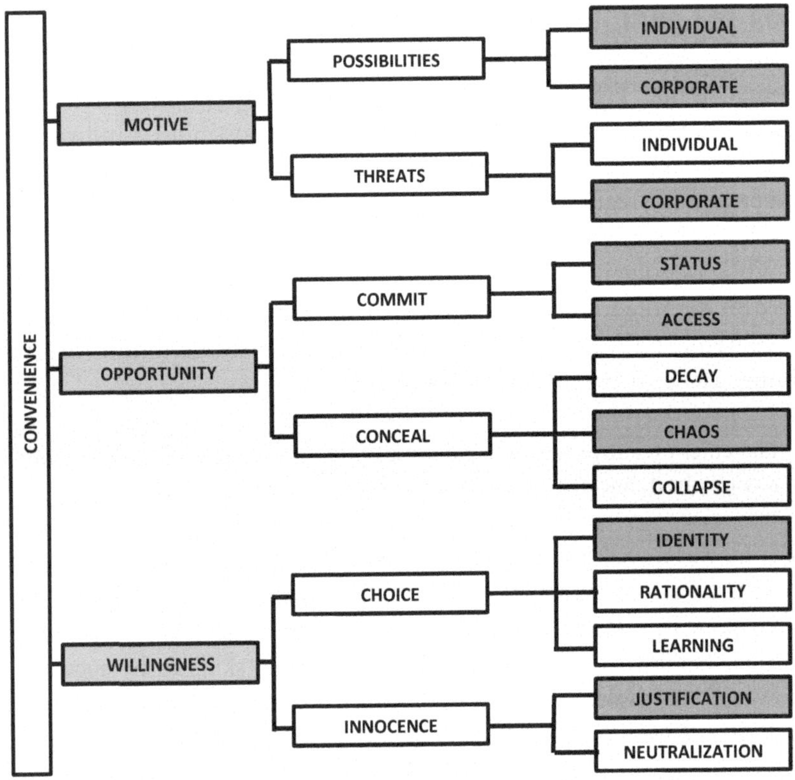

Fig. 12.2 Convenience themes in the case of Kraft & Kultur

company cover his private expenses. When the Ordfront board discovered his wrongdoing, Benulic got the choice of leaving voluntarily, or they would terminate him. The board considered reporting him to the police. However, since that would imply bad publicity, the board instead entered into an agreement with Benulic that he should repay his unjust gain.

Benulic (2018), on the other hand, has a slightly different version of what happened in his autobiography. He says he had to choose between leaving Ordfront and dismissal after dissatisfying top management, and that he was neither the first nor the last to have conflicts with them. Benulic felt that one of the reasons for the resignation was that the CEO of Ordfront got news of some rumors that employees at the company whispered at times that Benulic had been a better executive than he was. Benulic himself claims in his autobiography that he was not interested in the position, and that he was already mentally elsewhere.

After his time at Ordfront, he started what was to become the precursor to Kraft & Kultur, Hellefórs Kulturdistribution. However, this company went bankrupt already in 2001. The bankruptcy report shows that in the summer of 2000, the company bought book clubs with associated customer registers from Ordfront for the equivalent of two hundred thousand dollars. The bankruptcy report emphasizes that this obligation was initially Benulic's personal liability, but that he transferred it to the newly started company. Following the acquisition of clubs and registers, the company's financial situation deteriorated quickly, and the company tried to raise new capital without succeeding. Benulic (2018) tells in his book that Hellefórs started running out of money as they had spent a lot on redeeming member registers and customer orders from Ordfront. He does not mention that this was his personal debt. What he does, on the other hand, is to spend many pages reflecting on the desperation he felt when they started running out of money and bankruptcy could become a fact. He says that he spent a lot of time thinking about how to solve the situation. He says that at some point a Hells Angels member who offered to borrow him money contacted him. The next day another person invited him to a deserted cottage where the person offered him a loan. On the third day, the bank called him, telling him that they wanted to lend him four times the amount from the previous two. Benulic says he was doubtful about the first two and ended up refusing. The offer from the bank, however, was what made him most frightened. He would rather see the company go bankrupt than accept this loan. So Benulic (2018) says in his autobiography. However, the whole scene seems written in a way that makes it unclear if it actually happened or if this is something, he made up. He appears distant and strange, especially considering that he writes that borrowing from a bank is the scariest option, without presenting any reasons for it.

Benulic seems to have suffered from narcissistic identification with Kraft & Kultur since he created the business. Narcissistic identification is a special type of narcissism, where the offender sees little or no difference between self and the corporation. The company money is personal money that can be spent whatever way the narcissist prefers (Galvin et al., 2015). While narcissistic grandiosity and admiration belong to the motivational dimension of convenience theory, narcissistic empathy deficit belongs to the willingness dimension of convenience theory where the offender possesses a sense of entitlement (Nichol, 2019). The narcissistic offender shows unreasonable expectations to receive and obtain preferential treatments (Zvi and Elaad, 2018).

Kraft & Kultur was a dormant company from 1998 to 2001. The year 2002 was the first with real business. In 2001, however, several transactions took place with Helleförs Kulturdistribution in advance of their bankruptcy. Some transactions had no accounting documents at all. These transactions totaled the equivalent of one hundred thousand dollar. The board requested a statement from Benulic on these transactions for the next board meeting. Benulic never made a statement, and the case ended with it.

In the willingness dimension of convenience theory, Benulic was always ambivalent in accepting responsibility. On November 12, 2011, Benulic wrote in an email to the CEO at Troms Kraft that Kraft & Kultur had misleading numbers in accounts receivable. In the same email, he took on all the blame and wrote that nobody else had been involved. A few days later, he wrote a new email in which he resigned as general manager because of his health condition, but at the same time, he wanted to continue in the company in another function. Another couple of days after that he wrote a third email suggesting this time that he can take over the culture part of the company and the employees there (Ernst & Young, 2012).

Benulic later changed his explanation. During the trial in Södertörn (2015) district court, he denied that he had done anything wrong or punishable. Benulic claimed that he always tried to provide a true picture of Kraft & Kultur's finances. He blamed start-up problems. Benulic claimed that he had tried to make the accounting records for billable revenues as accurate as possible. Because he found that the accounting system provided misleading financial results, he manually changed the accounts to show the company's real economic volume. According to himself, he has thus done nothing wrong, since he only manually entered existing contracts and agreements in the accounts, and thus entered real expected income.

During interviews with fraud examiners from Ernst & Young (2012), Benulic said that he felt that he was not to blame for something going wrong. He named others to blame. Elite members can apply the blame game by misleading attribution to others (Eberly et al., 2011), or they can apply offender humor to distract attention from deviant behavior (Yam et al., 2018). Benulic blamed other executives at Kraft & Kultur for pricing electricity too low. He said during interviews that he relied on other executives' explanations and reports, and that if electricity deliveries subsequently proved to suffer from incorrect pricing, then Troms Kraft should hold other executives accountable.

Another example of misleading attribution in the interviews with Ernst & Young (2012) was when asked if there was any possibility that he had miscalculated. Benulic replied that he could not have been wrong. However, he pointed out that the supporting material, the excel files, that he got from others might have had errors from manipulation. Benulic's answer might be interpreted as meaning that he admits that something might have been wrong, but that he at the same time slides the responsibility for potential wrongdoing away from himself and over to other executives.

In his autobiography, Benulic (2018) stands completely unaware of the charges against him. He claims that even his lawyers do not understand at all why the prosecutor has indicted him. Benulic points out that he had no personal financial gain from the alleged wrongdoing. He spent much time delegating tasks to other people in the organization. He argues that it would not have been rational for him to manipulate accounts. He had also made it clear to the board of Kraft & Kultur and corporate management that he did not want to be part of the audit work. Benulic uses the last part of his autobiography to explain why it was not he, who performed the manipulation. It is thus obvious from reading his autobiography that Benulic does not take on any responsibility, admits no guilt, and argues that nobody should hold him accountable for any wrongdoing.

Benulic seems smart in a strange way. In some of the email examples cited by Ernst & Young (2012), it should normally be difficult for him to explain some formulations away. Phrases like "I can't conjure up the cheat every six months," "the auditors seem to buy all my explanations," and "triumph with them too, soon we can perform in Las Vegas with this act" are all pointing in the direction that he knew he did something that was at best in the gray zone of the law.

During the interviews with examiners, it appears that Benulic on several occasions has been sitting on information about Kraft & Kultur, which he chose not to share with the board. For example, Benulic was aware that Troms Kraft had prepared an electricity trading policy, and he was familiar with it. Nevertheless, he decided that Kraft & Kultur should not follow this and not price-hedge the fixed price contracts. While this convenience theme of Benulic deviating from official policy without reactions from anyone belongs in the opportunity dimension of convenience theory, his conviction of being right belongs in the willingness dimension. He gave clear instructions to employees at Kraft & Kultur that they could not talk to anyone in other parts of the Troms Kraft group about it. Since they were obedient, this phenomenon belongs in the opportunity dimension. Obedience in human behavior is a form of social influence receipt in which a person learns how to adhere to instructions or orders from a person of authority (Baird & Zelin, 2009; Hollow, 2014).

It is thus clear that Benulic deliberately chose to hide information about corporate deviance from the board of directors as well as the group as a whole. Another example is that Benulic, during an interview with examiners, acknowledged that he was aware of weaknesses in the basic documentation of the financial statements, and that he chose not to inform the board of this. In this instance, too, he chose to hide information from the board and the rest of the group.

Benulic admits in his autobiography that during the first years of Kraft & Kultur they engaged in misleading marketing. They marketed themselves as a company that operated within culture and environmental protection, while until 2006 the company offered only green energy in the alleged form of electricity from hydropower. This changed when they in 2006 decided to offer more environmentally friendly products, books, and films. Kraft & Kultur bought power from Nord Pool. That is, they bought energy from a common source, and they thus had no way of telling what proportion of the power they resold came from renewable sources such as waterfalls. The green energy they marketed might not have been as green as he claimed it to be (Benulic, 2018).

For a while, financial performance at Kraft & Kultur looked good. Margins in 2003 were high due to a currency gain. This foreign exchange gain was not clearly reflected in the accounts and was not reported to the board, as can been seen in the board minutes (Ernst & Young, 2012). It was probably a deliberate act by Benulic to make the revenues for that year look better than what reality would tell.

Elements of pride, patronizing attitude, and lack of sense of or need for belonging are some of the personal attributes of Benulic as an offender. In his autobiography, he says that in the first period at Hellefors Kulturdistribusjon, he worked with two others at their home before renting a small office space in what he calls a bad area in Stockholm. For a while, the three persons worked without pay. Benulic's (2018) formulations seem to indicate that he experienced a great deal of pride and high work ethics among the three because they worked hard despite poor working conditions. Furthermore, he says that at the beginning of the twenty-first century, it was new and unfamiliar to Swedish customers that they could freely choose electricity providers. He proudly states that Kraft & Kultur carried customers out of prisons where they seemed trapped, and then offered them a better deal. He believes that their initial marketing strategy is the reason why the company became one of Sweden's largest electricity providers a few years later.

In Benulic's testimony, Benulic reveals that he, according to himself, is the only one who can give a correct calculation of accrued income, and thus the correct presentation of the economy in Kraft & Kultur. According to Benulic, neither systems for accounting nor consultants for auditing were capable of generating correct numbers for expenses and revenues (Södertörn, 2015).

Benulic gives the impression that he has high thoughts about himself, also relative to other people, in his autobiography. He considers himself an important person, even after conviction and incarceration. He also gives the impression that he has little confidence in professionals such as engineers and lawyers, and the system in general. He thinks he has better knowledge and understanding than others do have. This is evident, for example, when he tells about an incident at the bank. He says he noticed that he was an important customer at the bank, which he does not think is strange considering the amounts they borrow. Evidence of his importance is the serving of coffee in a porcelain cup rather than a cardboard mug in the bank. In addition, he had the opportunity to review his personal finances during the meeting. However, he was not interested. He describes the bank adviser as a young woman wearing tailor-made clothing and pearl jewelry. He believes that she has not read more than a few lines about the funds she presents. He writes that he devised many questions that he did not ask, because he did not want to offend her. He says instead that he would rather put all his money under his mattress than buy funds, as he considers it a better choice considering the world economy. He says this in all seriousness.

Benulic (2018) also writes about several meetings with the chief financial officer at Troms Kraft concerning whether to fix interest rates or not, and he considers these meetings to be completely absurd. He ponders complex financial issues every day and fails to make decisions. Maybe his lack of financial training makes it difficult for him to understand the meaning of meetings. He wonders why they are important, and what is most appropriate to do. He does not trust that the bank adviser in tailor-made clothes can provide relevant information. He is even more puzzled about the woman who recommends a product she does not understand. He notices lack of knowledge among people who make recommendations, which they do not understand themselves.

Benulic is concerned with high work ethics and believes that many people lack it. For example, he feels relief when Hellefors Kulturdistribution goes bankrupt because he no longer had to deal with others' reluctance on a daily basis toward all demands to make an effort. He also complains that employees will not work overtime without pay (Benulic, 2018).

Benulic (2018) is concerned about what he perceives as people's care to fit in. Sweden is a country with a consensus culture. Already in his time as a journalist, he reacted to other journalists who were more concerned with fitting in than doing a good job and telling the important stories.

He describes in his autobiography several situations where he was not at all in agreement with other people. An example is when some representatives from Troms Kraft showed up in Stockholm, and Benulic joined them for dinner. He describes the representatives as seven little drunk dwarves.

An example that he is not interested in fitting in is from the first period at Kraft & Kultur when it was important for the company to sign new customers. He did not want to go. Instead, a young woman went, who he claims was interested in meeting new people, and who was interested in what the customers had to say.

He says he arrived at the office every day and was surprised to see all the strange people. He paused by the women only discussing silicone, while the men played video games in the lunchroom. He believes that too few read the books and magazines that the company published. He thinks that their hiring was a result of managers who did not want to hire anyone who could challenge their positions in the company. Other executives were only interested in talking about company cars for themselves, about summer resorts, and about where to go on the next vacation. He was thus not encouraged to talk to them either (Benulic, 2018).

Accounting Fraud

On November 11, 2011, a whistleblower detected that revenues, and thus financial results at Kraft & Kultur in Sweden, were unrealistically high for a number of years. As a result, there were items in the balance sheet that were incorrect. The mandate for the fraud investigation by Ernst & Young (2012) included the following issues:

- Decision support for the board at Kraft & Kultur. What kind of information did the auditors provide the board and corporate executives?
- Actions taken by the board at Kraft & Kultur. How did they handle information, and what kind of decision-making did take place?
- Information sharing. What quantity and quality of information arrived at board meetings from corporate management?

- Timing of fraud detection. Are there circumstances indicating that detection should have occurred at an earlier point in time?

Accounting manipulation is the misrepresentation of financial figures for income, costs, and assets. Financial statement fraud in terms of financial information manipulation is a distorted presentation, a misstatement of the financial performance and position of an enterprise. The purpose is to create a false impression of an organization's financial strength. Accounting manipulation creates an asymmetry of knowledge for readers of financial statements versus creators of financial statements. In many companies, internal controls of financial reporting have no priority (Gao & Zhang, 2019).

The external investigation by auditing firm Ernst & Young (2012) found strong indications that revenues was overstated as a result of many years of manipulations. The fraud examination report suggests that CEO Benulic was the main architect of the fraudulent financial reporting, with the help of other executive members of management. Benulic was one of the founders of Kraft & Kultur and was the chief executive in the company from the beginning. He obviously wanted to perform well in his job, and accounting fraud was a convenient avenue for window dressing.

Accounting is of great importance to any organization. A number of employees record a large number and variety of transactions each day. The appearance should reflect the real financial situation at any time to enable relevant decision-making for internal and external stakeholders. Accounting needs to emerge in accordance with the rules, principles, and standards as well as best practices.

The financial management function at Kraft & Kultur never had a strong position. Benulic had the overall responsibility for financial matters and administration from the start in 2001. In 2003, the company board decided to let Benulic manage all company finances and preparations of management accounting. This responsibility was only supposed to last until professional reporting routines were working. However, Benulic remained in charge of financial reporting (Nødset, 2012).

Corporate investigators draw the following conclusions regarding the opportunity structure for Benulic to commit accounting fraud (Ernst & Young, 2012: 7):

- Despite several comments from Kraft & Kultur's auditor, the situation with partly weak internal control continued and especially until 2008.
- The administration was small and person dependent, and the chief executive had seemingly much power and influence.
- Financial reporting from the administration and Kraft & Kultur throughout the period has been substandard, with some improvements starting 2008.
- The company introduced in 2008 a unified corporate reporting, which also included Kraft & Kultur. Despite more visible reporting of outstanding income, which was increasing, and especially since 2009, reviewers cannot see that the board at Kraft & Kultur before the year 2011 implemented special actions to reduce the balance amount.
- Through the years, the chief executive had contact and meetings with auditors concerning all questions related to accounting and financial reporting, including outstanding income.

Board members at Kraft & Kultur had knowledge of the energy industry, but no competence in management and financial matters (Ernst & Young, 2012). They claimed that they were unable to uncover the fraud, as they felt being victims of manipulation by Benulic (Nødset, 2012). While board members lacked knowledge of financial matters, Benulic as chief executive had complete control over accounting and auditing. In addition, Benulic was "charismatic, authoritarian, and someone who taught himself the energy business in a very short time" (Ernst & Young, 2012: 28).

In conclusion, the purpose of this chapter was to illustrate the application of convenience theory to the case of a convicted white-collar offender. The case study has documented how the structural model of convenience theory can be applied to a specific white-collar offender to identify relevant convenience themes that can explain the occurrence of crime. The case study was based on archival analysis of Swedish court documents, fraud investigation reports, an autobiography by the former

CEO, and media reports. His motive seems dominated by individual and corporate possibilities as well as corporate threats. His opportunity seems dominated by status, access, and chaos. His willingness seems dominated by identity and justification.

Since the theory of convenience is an emerging perspective on white-collar and corporate crime, there are several avenues for future research in terms of both theoretical and empirical studies. Identification of theoretical limitations can open up for new convenience themes. Empirical studies of offenders versus non-offenders can enable testing of theoretical assumptions in convenience theory.

Bibliography

Adler, P. S., & Kwon, S. W. (2002). Social capital: Prospects for a new concept. *Academy of Management Review, 27*(1), 17–40.

Agnew, R. (2014). Social concern and crime: Moving beyond the assumption of simple self-interest. *Criminology, 52*(1), 1–32.

Alon, A., Mennicken, A., & Samsonova-Taddei, A. (2019). Dynamics and limits of regulatory privatization: Reorganizing audit oversight in Russia. *Organization Studies, 40*(8), 1217–1240.

Baird, J. E., & Zelin, R. C. (2009). An examination of the impact of obedience pressure on perceptions of fraudulent acts and the likelihood of committing occupational fraud. *Journal of Forensic Studies in Accounting and Business, 1*(1), 1–14.

Bell, C. M., & Khoury, C. (2016). Organizational powerlessness, dehumanization, and gendered effects of procedural justice. *Journal of Managerial Psychology, 31*, 570–585.

Bell, E., Bryman, A., & Harley, B. (2018). *Business research methods* (2nd ed.). New York, NY: Oxford University Press.

Benartzi, S., Beshears, J., Milkman, K. L., Sunstein, C. R., Thaler, R. H., Shankar, M., et al. (2017). Should governments invest more in nudging? *Psychological Science, 28*(8), 1041–1055.

Benson, M. L., & Simpson, S. S. (2018). *White-Collar crime: An opportunity perspective* (3rd ed.). New York, NY: Routledge.

Benulic, B. (2018). *Inte mitt krig* [Not my war]. Sweden: Cultura Aetatis Publishing.

Bernburg, J. G., Krohn, M. D., & Rivera, C. (2006). Official labeling, criminal embeddedness, and subsequent delinquency. *Journal of Research in Crime and Delinquency, 43*(1), 67–88.

Blickle, G., Schlegel, A., Fassbender, P., & Klein, U. (2006). Some personality correlates of business white-collar crime. *Applied Psychology: An International Review, 55*(2), 220–233.

Bosse, D. A., & Phillips, R. A. (2016). Agency theory and bounded self-interest. *Academy of Management Review, 41*(2), 276–297.

Braaten, C. N., & Vaughn, M. S. (2019). Convenience theory of cryptocurrency crime: A content analysis of U.S. federal court decisions. *Deviant Behavior*, published online https://doi.org/10.1080/01639625.2019.1706706.

Bussmann, K. D., Niemeczek, A., & Vockrodt, M. (2018). Company culture and prevention of corruption in Germany. *China and Russia, European Journal of Criminology, 15*(3), 255–277.

Chan, F., & Gibbs, C. (2020). Integrated theories of white-collar and corporate crime. In M. L. Rorie, (Ed.), *The handbook of white-collar crime*. Hoboken, NJ: Wiley & Sons, chapter 13, pages 191–208.

Chang, J. J., Lu, H. C., & Chen, M. (2005). Organized crime or individual crime? *Endogeneous size of a criminal organization and the optimal law enforcement, Economic Inquiry, 43*(3), 661–675.

Chatterjee, A., & Pollock, T. G. (2017). Master of puppets: How narcissistic CEOs construct their professional worlds. *Academy of Management Review, 42*(4), 703–725.

Cohen, L. E., & Felson, M. (1979). Social change and crime rate trends: A routine activity approach. *American Sociological Review, 44*, 588–608.

Comey, J. B. (2009). Go directly to prison: White collar sentencing after the Sarbanes-Oxley act. *Harvard Law Review, 122*, 1728–1749.

Craig, J. M., & Piquero, N. L. (2017). Sensational offending: An application of sensation seeking to white-collar and conventional crimes. *Crime & Delinquency, 63*(11), 1363–1382.

Cressey, D. (1972). *Criminal organization: Its elementary forms*. New York, NY: Harper and Row.

Crosina, E., & Pratt, M. G. (2019). Toward a model of organizational mourning: The case of former Lehman Brothers bankers. *Academy of Management Journal, 62*(1), 66–98.

Davidson, R. H., Dey, A., & Smith, A. J. (2019). CEO materialism and corporate social responsibility. *The Accounting Review, 94*(1), 101–126.

Dearden, T. E., & Gottschalk, P. (2020). Gender and white-collar crime: Convenience in target selection. *Deviant Behavior*, published online doi 10.1080/01639625.2020.1756428.

Deloitte. (2017). *Investigation Report*. Independent Investigation Committee, by global auditing firm Deloitte, published June 10, Ito, T., Sato, K. and Nishimura, K., https://www.fujifilmholdings.com/en/pdf/investors/finance/materials/ff_irdata_investigation_001e.pdf, downloaded September 8, 2018.

Duriau, V. J., Reger, R. K., & Pfarrer, M. D. (2007). A content analysis of the content analysis literature in organization studies: Research themes, data sources, and methodological refinements. *Organizational Research Methods, 10*(1), 5–34.

Eberly, M. B., Holley, E. C., Johnson, M. D., & Mitchell, T. R. (2011). Beyond internal and external: A dyadic theory of relational attributions. *Academy of Management Review, 36*(4), 731–753.

Engdahl, O. (2015). White-collar crime and first-time adult-onset offending: Explorations in the concept of negative life events as turning points. *International Journal of Law, Crime and Justice, 43*(1), 1–16.

Ernst & Young. (2012). *Troms Kraft AS – Gransking av Kraft & Kultur i Sverige AB* [Troms Kraft Inc.—Investigation of Kraft & Kultur in Sweden Inc.]. Global auditing firm Ernst & Young, Oslo, Norway.

Farquhar, J. D., & Rowley, J. (2009). Convenience: a services perspective. *Marketing Theory, 9*(4), 425–438.

Ferraro, F., Pfeffer, J., & Sutton, R. I. (2005). Economics language and assumptions: How theories can become self-fulfilling. *Academy of Management Review, 30*(1), 8–24.

Galvin, B. M., Lange, D., & Ashforth, B. E. (2015). Narcissistic organizational identification: Seeing oneself as central to the organization's identity. *Academy of Management Review, 40*(2), 163–181.

Gamache, D. L., & McNamara, G. (2019). Responding to bad press: How CEO temporal focus influences the sensitivity to negative media coverage of acquisitions. *Academy of Management Journal, 62*(3), 918–943.

Gao, P., & Zhang, G. (2019). Accounting manipulation, peer pressure, and internal control. *The Accounting Review, 94*(1), 127–151.

Goldstraw-White, J. (2012). *White-collar crime: Accounts of offending behavior*. London, UK: Palgrave Macmillan.

Goncharov, I., & Peter, C. D. (2019). Does reporting transparency affect industry coordination? *Evidence from the duration of international cartels, The Accounting Review, 94*(3), 149–175.

Gottfredson, M. R., & Hirschi, T. (1990). *A general theory of crime*. Stanford: Stanford University Press, CA.
Hamish, M. (2016, December 24). SFO closes Fuji Xerox probe. *The Press*. https://infoweb.newsbank.com/apps/news/document-view?p=AWNB&t=&sort=YMD_date%3AD&maxresults=20&f=advanced&val-base-0=fuji%20xerox&fld-base-0=alltext&bln-base-2=and&val-base-2=new%20zealand&fld-base-2=alltext&bln-base-3=and&val-base-3=2016&fld-base-3=YMD_date&docref=news/16175F5693A0BFF0.
Hansen, L. L. (2020). Review of the book "convenience triangle in white-collar crime: Case studies of fraud examinations". *ChoiceConnect*, *57*(5). Middletown, CT: Association of College and Research Libraries.
Henderson, J. (2016). Neil Whittaker exits as local Fuji Xerox MD in shock departure. *ARN Net*, published May 18, https://www.arnnet.com.au/article/600045/shock-departure-neil-whittaker-exits-local-fuji-xerox-md/.
Higgins, E. T. (1997). Beyond pleasure and pain. *American Psychologist, 52,* 1280–1300.
Hoffmann, J. P. (2002). A contextual analysis of differential association, social control, and strain theories of delinquency. *Social Forces, 81*(3), 753–785.
Hollow, M. (2014). Money, morals and motives. *Journal of Financial Crime, 21*(2), 174–190.
Hsieh, H., & Shannon, S. E. (2005). Three approaches to qualitative content analysis. *Qualitative Health Research, 15*(9), 1277–1288.
Huang, L., & Knight, A. P. (2017). Resources and relationships in entrepreneurship: An exchange theory of the development and effects of the entrepreneur-investor relationship. *Academy of Management Review, 42*(1), 80–102.
Huisman, W., & Erp, J. (2013). Opportunities for environmental crime. *British Journal of Criminology, 53,* 1178–1200.
Hurley, P. J., Mayhew, B. W., & Obermire, K. M. (2019). Realigning auditors' accountability: Experimental evidence. *The Accounting Review, 94*(3), 233–250.
Jonnergård, K., Stafsudd, A., & Elg, U. (2010). Performance evaluations as gender barriers in professional organizations: A study of auditing firms, *Gender, Work and Organization, 17*(6), 721–747.
Karim, K. E., & Siegel, P. H. (1998). A signal detection theory approach to analyzing the efficiency and effectiveness of auditing to detect management fraud. *Managerial Auditing Journal, 13*(6), 367–375.

Kawasaki, T. (2020). Review of comparative studies on white-collar and corporate crime. In M. L. Rorie (Ed.), *The handbook of white-collar crime*. Hoboken, NJ: Wiley & Sons, chapter 27, pages 437–447.

Keil, M., Tiwana, A., Sainsbury, R., & Sneha, S. (2010). Toward a theory of whistleblowing intentions: A benefit-cost differential perspective. *Decision Sciences, 41*(4), 787–812.

Krippendorff, K. (1980). *Content analysis: An introduction to its methodology*. Beverly Hills, CA: Sage.

Langton, L., & Piquero, N. L. (2007). Can general strain theory explain white-collar crime? *A preliminary investigation of the relationship between strain and select white-collar offenses, Journal of Criminal Justice, 35*(1), 1–15.

Lehman, D. W., Cooil, B., & Ramanujam, R. (2019). The effects of rule complexity on organizational noncompliance and remediation: Evidence from restaurant health inspections. *Journal of Management*, published online, 1–33. https://doi.org/10.1177/0149206319842262.

Leigh, A. C., Foote, D. A., Clark, W. R., & Lewis, J. L. (2010). Equity sensitivity: A triadic measure and outcome/input perspectives. *Journal of Managerial Issues, 22*(3), 286–305.

Leonard, W. N., & Weber, M. G. (1970). Automakers and dealers: A study of criminogenic market forces. *Law & Society Review, 4*(3), 407–424.

Mai, H. T. X., & Olsen, S. O. (2016). Consumer participation in self-production: The role of control mechanisms, convenience orientation, and moral obligation. *Journal of Marketing Theory and Practice, 24*(2), 209–223.

Maslow, A. H. (1943). A theory of human motivation. *Psychological Review, 50*(4), 370–396.

Mawritz, M. B., Greenbaum, R. L., Butts, M. M., & Graham, K. A. (2017). I just can't control myself: A self-regulation perspective on the abuse of deviant employees. *Academy of Management Journal, 60*(4), 1482–1503.

McClanahan, B., & South, N. (2020). 'All Knowledge begins with the senses': Towards a sensory criminology. *British Journal of Criminology, 60*, 3–23.

McClelland, P. L., Liang, X., & Barker, V. L. (2010). CEO commitment to the status quo: Replication and extension using content analysis. *Journal of Management, 36*(5), 1251–1277.

Mingus, W., & Burchfield, K. B. (2012). From prison to integration: Applying modified labeling theory to sex offenders. *Criminal Justice Studies, 25*(1), 97–109.

Mo, M. L. (2018). Nå må Boris Benulic sone i fengsel [Now Boris Benulic must be jailed], online newspaper *iTromsø*, www.itromso.no, published May 8.

Mohliver, A. (2019). How misconduct spreads: Auditors' role in the diffusion of stock-option backdating. *Administrative Science Quarterly, 64*(2), 310–336.

Naylor, R. T. (2003). Towards a general theory of profit-driven crimes. *British Journal of Criminology, 43,* 81–101.

Nergaard, L. L. (2013). *Sammendrag av granskingsrapport – Troms Kraft AS* [Summary of investigation report—Troms Kraft Inc.]. Norscan Partners, Norway.

Nichol, J. E. (2019). The effects of contract framing on misconduct and entitlement. *The Accounting Review, 94*(3), 329–344.

Nielsen, M. K. (2019). New report concludes: How Britta Nielsen could swindle for DKK 120 million for 25 years, Danish public broadcasting *Danmarks Radio*, www.dr.no, published March 1.

Nielsen, R. P. (2003). Corruption networks and implications for ethical corruption reform. *Journal of Business Ethics, 42*(2), 125-149.

Nødset, I. F. (2012, December 19). *Earnings management environment with the case of Troms Kraft* (Master of science thesis). Norwegian School of Economics, Bergen, Norway.

Obodaru, O. (2017). Forgone, but not forgotten: Toward a theory of forgone professional identities. *Academy of Management Journal, 60*(2), 523–553.

Olafsen, A. H., Niemiec, C. P., Halvari, H., Deci, E. L., & Williams, G. C. (2017). On the dark side of work: a longitudinal analysis using self-determination theory. *European Journal of Work and Organizational Psychology, 26*(2), 275–285.

Patel, P. C., & Cooper, D. (2014). Structural power equality between family and nonfamily TMT members and the performance of family firms. *Academy of Management Journal, 57*(6), 1624–1649.

Patrucco, A. S., Luzzini, D., & Ronchi, S. (2017). Research perspectives on public procurement: Content analysis of 14 years of publications in the Journal of Public Procurement. *Journal of Public Procurement, 16*(2), 229–269.

Petrocelli, M., Piquero, A. R., & Smith, M. R. (2003). Conflict theory and racial profiling: An empirical analysis of police traffic stop data. *Journal of Criminal Justice, 31*(1), 1–11.

Piquero, N. L. (2012). The only thing we have to fear is fear itself: Investigating the relationship between fear of falling and white-collar crime. *Crime and Delinquency, 58*(3), 362–379.

Piquero, N. L. (2018). White-collar crime is crime: Victims hurt just the same. *Criminology & Public Policy, 17*(3), 595–600.

Pontell, H. N., Black, W. K., & Geis, G. (2014). Too big to fail, too powerful to jail? On the absence of criminal prosecutions after the 2008 financial meltdown. *Crime, Law and Social Change, 61*(1), 1–13.

Pratt, T. C., & Cullen, F. T. (2005). Assessing macro-level predictors and theories of crime: A meta-analysis. *Crime and Justice, 32,* 373–450.

Qiu, B., & Slezak, S. L. (2019). The equilibrium relationships between performance-based pay, performance, and the commission and detection of fraudulent misreporting. *The Accounting Review, 94*(2), 325–356.

Ramoglou, S., & Tsang, E. W. K. (2016). A realist perspective of entrepreneurship: Opportunities as propensities. *Academy of Management Review, 41,* 410–434.

Rodriguez, P., Uhlenbruck, K., & Eden, L. (2005). Government corruption and the entry strategies of multinationals. *Academy of Management Review, 30*(2), 383–396.

Sampson, R. J., & Laub, J. H. (1993). *Crime in the making: Pathways and turning points through life*. Cambridge, MA: Harvard University Press.

Saunders, M., Lewis, P., & Thornhill, A. (2007). *Research methods for business students* (5th ed.). London, UK: Pearson Education.

Schnatterly, K., Gangloff, K. A., & Tuschke, A. (2018). CEO wrongdoing: A review of pressure, opportunity, and rationalization. *Journal of Management, 44*(6), 2405–2432.

Schoepfer, A., & Piquero, N. L. (2006). Exploring white-collar crime and the American dream: A partial test of institutional anomie theory. *Journal of Criminal Justice, 34*(3), 227–235.

Schoultz, I., & Flyghed, J. (2020a). Denials and confessions: An analysis of the temporalization of neutralizations of corporate crime, *International Journal of Law, Crime and Justice*, published online doi.org/10.1016/j.ijlcj.2020.100389.

Schoultz, I., & Flyghed, J. (2020b). "We have been thrown under the bus": Corporate versus individual defense mechanisms against transnational corporate bribery charges. *Journal of White Collar and Corporate Crime*, published online doi: 10.1177/2631309x20911883.

Stadler, W. A., Benson, M. L., & Cullen, E. T. (2013). Revisiting the special sensitivity hypothesis: The prison experience of white-collar inmates. *Justice Quarterly, 30*(6), 1090–1114.

Sundström, M., & Radon, A. (2015). Utilizing the concept of convenience as a business opportunity in emerging markets. *Organizations and Markets in Emerging Economies, 6*(2), 7–21.

Sutherland, E. H. (1939). White-collar criminality. *American Sociological Review, 5*(1), 1–12.
Sutherland, E. H. (1983). *White collar crime—The Uncut Version*. New Haven, CT: Yale University Press.
Svea. (2017). *Court sentence dated November 29 in the case 19590829-0017*. Svea court of appeal, Stockholm, Sweden.
Sykes, G. M., & Matza, D. (1957). Techniques of neutralization: A theory of delinquency. *American Sociological Review, 22*(6), 664–670.
Södertörn. (2015). *Court sentence dated August 7 in the case B 5582-12*, Södertörn district court, Stockholm, Sweden.
Vasiu, V. I., & Podgor, E. S. (2019, July). Organizational opportunity and deviant behavior: Convenience in white-collar crime. *Criminal Law and Criminal Justice Books*. Rutgers: The State University of New Jersey. www.clcjbooks.rutgers.edu.
Victor, B., & Cullen, J. B. (1988). The organizational bases of ethical work climates. *Administrative Science Quarterly, 33*, 101–125.
Weick, K. E. (1995). What theory is not, theorizing is. *Administrative Science Quarterly, 40*, 385–390.
Welsh, D. T., Ordonez, L. D., Snyder, D. G., & Christian, M. S. (2014). The slippery slope: How small ethical transgressions pave the way for larger future transgressions. *Journal of Applied Psychology, 100*(1), 114–127.
Wikstrom, P. O. H., Mann, R. P., & Hardie, B. (2018). Young people's differential vulnerability to criminogenic exposure: Bridging the gap between people- and place-oriented approaches in the study of crime causation. *European Journal of Criminology, 15*(1), 10–31.
Yam, K. C., Christian, M. S., Wei, W., Liao, Z., & Nai, J. (2018). The mixed blessing of leader sense of humor: Examining costs and benefits. *Academy of Management Journal, 61*(1), 348–369.
Zvi, L., & Elaad, E. (2018). Correlates of narcissism, self-reported lies, and self-assessed abilities to tell and detect lies, tell truths, and believe others. *Journal of Investigative Psychology and Offender Profiling, 15*, 271–286.

13
Conclusion

Policing environmental crime in terms of investigation and prosecution remains a complicated issue. Different from traditional street crime where offense, offender, and victim are identifiable and consistent with laws that prescribe punishment, environmental wrongdoing result in harm that soemtimes is illegal and sometimes not illegal. As emphasized in the last chapter, research needs to address how environmental crime is conceptualized and how criminalization processes develop. Certain mechanisms are interesting to study, which cause some of the harms from environmental wrongdoing to be legal while others are criminalized. Green criminology has become an important and expanding field in criminology, particularly within the perspective of critical criminology, which focuses on challenging traditional understandings and uncovering false beliefs about criminal justice.

The case study of Vest Tank illustrates that detection of environmental crime was dependent on an explosion that caused harm to local people. There was an offense of illegal waste disposal, there was an offender in terms of two executives and an expert, and there was a victim in terms of harm to local people. Therefore, the case was not that difficult to investigate and prosecute for final convictions.

P. Gottschalk, *Detecting and Investigating Environmental Crime*,
https://doi.org/10.1007/978-3-030-74184-6_13

More difficult, however, is the Tjøme case to investigate and prosecute as environmental crime. Such difficulty is the likely cause of Norwegian police first looking for traditional white-collar crime in terms of corruption. Økokrim chose to start the Tjøme case by charging and indicting a municipal employee and an architect for being the bribed and being the briber. The court hearings were scheduled for January 2021 with the two defendants. Later, the rich and mighty on the shoreline could be prosecuted and end up as defendants in court for environmental crime, probably in 2022.

While Økokrim chose this policing strategy of first corruption and then environmental harm in the Tjøme case in 2019, the final outcome of the Grimstad case in 2020 made the police more confident to prosecute environmental offenders. While the rich and mighty defendant Madsen brought her case all the way from the district court via a court of appeals to the Norwegian Supreme Court, she still ended up with a jail sentence in 2020. This kind of encouragement made Økokrim ask the councilor in the municipality of Færder (formerly Tjøme) to document some obvious incidents of wrongdoing by mansion owners on the shoreline.

As illustrated and exemplified in this book, the theory of convenience provides an explanation for environmental crime occurrences. A stronger motive, a simpler opportunity, and a higher willingness all contribute to the convenience of crime. The attractiveness of a mansion on the shoreline, where you can walk out of bed into a morning bath in the sea, where you can have your sailingboat ready, and where you can enjoy your wine or beer while watching the sun going down, is obvious. The availability of an architect with a network and an attorney with skills within substance defense, information control, and symbol defense represents an attractive opportunity structure. The perceived justification by having paid for it all, having caused no injury to people, and disagreeing with the hundred-meter ban, creates willingness for deviant behavior.

Policing environmental crime is not only a matter of detecting and prosecuting offenders. Policing is also concerned with prevention of offenses. Prevention tactics can result from a study of convenience theory where the objective is to reduce the convenience of white-collar crime in terms of motive, opportunity, and willingness.

In the motive dimension, we find that the American dream does not dominate Norwegian culture. Not that many Norwegians will agree that the greatest thing you can experience in life is to become rich. Meaningful work, knowledge, competence, friendship, family, and health are all issues that may be more important. By emphasizing the core values in the Norwegian culture, the motive for environmental crime might become weaker.

In the opportunity dimension, the best preventive measure is probably an increase in the subjective detection risk. If you believe that you will get caught, you do not do it. The subjectivity varies with individuals in the same situation where the objective detection risk is equal for all. It is a matter of perception that can be influenced by control mechanisms, whistleblowing routines, and deterrence from knowledge of convicted offenders. The preventive outcome of the Madsen conviction in Grimstad and of the indictments on Tjøme seem already visible in Norway. You cannot simply run away by paying a fine, and you cannot trust your attorney(s) to cause dismissal of indictments against you anymore.

In the willingness dimension, the rich and mighty have claimed entitlement to privileges that justify wrongdoing and remove feelings of guilt. However, as more people condemn their behavior out of environmental protection perspectives as well as climate issues, potential offenders become more isolated in their willingness thinking. Environmental protection is an issue that most Norwegians support. While Norwegians financially support preservation of rain forests in other parts of the world, and the wolf is allowed to live although it kills sheep, then the same Norwegians may find it harder to destroy and privatize the shoreline in the future.

Criminology is here different from law. While nobody can be convicted if there is no law sanctioning what they have done, everyone can be condemned for wrongdoing. In criminology, a crime has two characteristics. First, people find it wrong to do. Second, the wrongdoing deserves punishment. If these two characteristics are not present, people will not consider it crime. Environmental wrongdoing has climbed in the perception of seriousness in the public mind, where nobody objects anymore prosecution and incarceration. It is worth a thought experiment to consider whether Madsen in Grimstad would be prosecuted in the 1960s in Norway. At that time, a fine would probably be the only acceptable sanction in terms of what people found wrong to do, and what kind of punishment was required. Today, it is acceptable to convict somebody to prison for environmental crime in Norway. The two executives, the consultant, and the summerhouse owners, are examples of this trend of considering environmental crime more serious today compared to the past. This historical perspective is yet another reason to believe that Økokrim may be successful in prosecuting the rich and mighty in the beach zone on the island of Tjøme when the evidence is ready for it.

The case study of the municipality of Tjøme in this book has emphasized the role of whistleblower Tom Ravndal. He lived in the city of Tønsberg not far from the island of Tjøme where he had his own cabin. In the spring of 2020, he decided to move to his cabin and make it his permeant residence. He then became a citizen of Færder municipality. When politicians in the municipality noticed that he had moved to Færder, they nominated him for the community award:

> We will nominate Tom Ravndal for Færder Municipality's Community Prize 2020. He has carried out a unique and very extensive work when it comes to uncovering and informing about illegal construction along the coast, a work that is of great importance to the local community and the population in Færder. Tom Ravndal's selfless efforts have been crucial for the municipality to be able to safeguard the beach zone and the population's rights when it comes to access to our fantastic coastal nature.
>
> The access to the beach zone is of invaluable benefit to the population. Access to the coast is essential in terms of health-promoting activity and positive experiences for all groups of our local community. Our municipality's unique coastal nature is a national treasure that our municipality has an important responsibility to preserve. Protecting the coastal nature and the people's access to it promotes inclusion and a good local environment to live in. Not least, access to the coast is important for the growing up of children in our municipality, where they grow up with the beautiful coastal nature as an important part of their surroundings. This means a lot in terms of positive and health-promoting experiences and a positive relationship to the home municipality. For the locals, summerhouse guests as well as other visitors, access to the beach zone is essential

> for well-being, quality of life, and the continuation of our culture and our traditions.
>
> The uncovering of illegalities in the beach zone on Tjøme island, largely made possible by the work carried out by Tom Ravndal, has significantly contributed to preventing the weakening of the protection of the beach zone and the right of public access that has characterized the coast of Færder through illegal construction. Tom Ravndal's efforts have been a key element in connection with Færder being a municipality that is now focusing on illegality follow-up and measures to safeguard the people's rights to travel in rural areas along the coast. Tom Ravndal's efforts to safeguard Færder's coastline are also of great importance when it comes to the intrinsic value of coastal nature and its significance for biodiversity.

One of the few opponents to the nomination of Tom Ravndal detected that Ravndal had just recently moved to Færder. According to the statutes, the award winner had to be an inhabitant in the municipality for at least one year. The expectation and obvious prediction was that Ravndal would be awarded Færder Municipality's Community Prize 2021.

Index

P. Gottschalk, *Detecting and Investigating Environmental Crime*,
https://doi.org/10.1007/978-3-030-74184-6

The manufacturer's authorised representative in the EU is Springer Nature Customer Service Centre GmbH, Europaplatz 3, 69115 Heidelberg, Germany. If you have any concerns regarding our products, please contact ProductSafety@springernature.com

Printed and bound by CPI Group (UK) Ltd, Croydon, CR0 4YY
22/09/2026
02221066-0007